Electrochemical Modeling in the Context of Production of Lithium-based Batteries

Von der Fakultät für Maschinenbau
der Technischen Universität Carolo-Wilhelmina zu Braunschweig

zur Erlangung der Würde

eines Doktor-Ingenieurs (Dr.-Ing.)

genehmigte Dissertation

von: Vincent Laue, M.Sc.
geboren in (Geburtsort): Braunschweig

eingereicht am: 16.01.2020
mündliche Prüfung am: 15.06.2020

Vorsitz:

Prof. Dr.-Ing. Arno Kwade

Gutachter:

Prof. Dr.-Ing. Ulrike Krewer
Prof. Dr.-Ing. Richard Hanke-Rauschenbach

2021

Bibliografische Information der Deutschen Nationalbibliothek

Die Deutsche Nationalbibliothek verzeichnet diese Publikation in der
Deutschen Nationalbibliografie; detaillierte bibliografische Daten sind
im Internet über http://dnb.d-nb.de abrufbar.

ISBN 978-3-8325-5241-1

Logos Verlag Berlin GmbH
Georg-Knorr-Str. 4, Geb. 10,
D-12681 Berlin
Germany

Tel.:   +49 (0)30 / 42 85 10 90
Fax: +49 (0)30 / 42 85 10 92
http://www.logos-verlag.de

Contents

Abstract

Lithium ion batteries have outperformed other battery technologies for decades. Further improvement requires enhancement of the materials, as well as optimization of electrode composition and processing. To enable application of electrochemical models in design optimization and battery production, this dissertation addresses the challenges and applications of battery models. Therefor, an electrochemical pseudo-2D battery model is parameterized assessing its practical model identifiability. Applying this model an uncertainty quantification is carried out to reveal relevant processes. Further, a 3D micro structure model is developed to address the influence of spatial effects, for instance due to the distribution of electron conducting additives. Eventually, the electrochemical and the micro structure model are coupled applying empirical surrogate models. The feasibility of the extended model to simulate the calendering influence is shown. The results show a need for three-electrode measurements with static and dynamic electrochemical tests to enable model identifiability. Further, the uncertainty quantification revealed the nonlinearity and C-rate dependency of the uncertainty propagation. Eventually, micro structure simulations revealed optimization potentials of all-solid-state electrodes through knowledge-based design of mixing routines to maximize electrical and ionic conductivity. The introduced models and the derived results contribute to the simulations-based optimization of battery production and the enhancement of battery performance.

Kurzfassung

Lithiumionen-Batterien finden Verwendung in einer Vielzahl technischer Anwendungen von Mobiltelefonen hin zu Elektrofahrzeugen und stationären Energiespeichern zur Netzstabilisierung. Mit der Verbreitung der Lithiumionen-Batterie steigen Anforderungen an Leistung und Qualität der Batterien. Zudem müssen Produktionskosten und der Einfluss der Batterieproduktion auf die Umwelt reduziert werden. Angesichts dieser Herausforderungen rückt die Batterieproduktion selbst in den Fokus der Wissenschaft und auch der Einsatz von elektrochemischen Modellen in der Produktion wird verstärkt diskutiert.

Entsprechend dieser Ausgangslage ist das Ziel dieser Dissertation Möglichkeiten und Limitierungen der elektrochemischen Modellierung im Kontext der Produktion von Lithium-basierten Batterien aufzuzeigen. Anschließend sollen die Modelle weiterentwickelt werden, um ihre Einsatzmöglichkeiten in der Forschung, aber auch in der Produktion zu erweitern.

Neben den Modellgleichungen sind die Parameter ein essentieller Bestandteil eines Modells, wobei die Parametrierung nicht trivial ist, da ein Teil der Parameter nicht, oder nur schwer experimentell bestimmt werden kann. Daher wurde zu Beginn dieser Arbeit das weitverbreitete pseudo-zweidimensionale Doyle-Newman-Modell auf eine Identifizierbarkeit anhand von Standardexperimenten untersucht. Ergebnis war eine weitgehende Unidentifizierbarkeit an C-Raten-Tests, wobei diese Limitierung mit zusätzlichen Impedanzdaten und der Erfassung von Halbzellpotentialen überwunden werden konnte. Das parametrierte Modell wurde verwendet, um die Fortpflanzung von Unsicherheiten zwischen Zwischenprodukteigenschaften und Produkteigenschaften in der Batterieproduktion zu quantifizieren. Dies ermöglichte die Identifikation relevanterer Prozesse in der Zelle, sowie in der Produktion. Um den Gültigkeitsbereich des Batteriemodells zu erweitern und so z.B. eine Optimierung des Kalandriergrades in der Produktion zu ermöglichen, wurde ein dreidimensionales Mikrostrukturmodell entwickelt, dass die drei Hauptbestandteile der Elektroden (Aktivmaterial, Elektrolyt, Leitadditive/Binder) und deren räumliche Verteilung berücksichtigt. Dieses Modell wurde einerseits dazu genutzt, Ansätze zur Optimierung der Prozessierung von Festelektrolyt-Elektroden zu identifizieren und zu bewerten, andererseits wurden empirische Surrogatmodelle entwickelt, die das bestehende elektrochemische Batteriemodell um die komplexen Mikrostruktureinflüsse aus dem Mikrostrukturmodell erweitern. Diese Modellerweiterung steigert den Gültigkeitsbereich des Modells hinsichtlich der Elektrodenzusammensetzung signifikant, ohne die Rechenzeit nennenswert zu erhöhen.

Zusammenfassend tragen die Ergebnisse dieser Dissertation dazu bei, elektrochemische Modelle in der Batterieproduktion zu etablieren. Hinsichtlich der Identifizierbarkeit von Modellen wurden Handlungsempfehlungen für zukünftige Parametrierungen abgeleitet, und in der elektrochemischen Modellierung, wie auch in der Mikrostrukturmodellierung wurde der Einfluss der festen Leitadditive erstmals detailliert modelbasiert quantifiziert. Diese Modelle, sowie das gewonnene physikalische Verständnis über interne Prozesse der Batterie können perspektivisch dazu dienen, Batterien weiter zu optimieren und den Einfluss der Produktion auf die Umwelt, sowie die Kosten durch Vermeidung von Ausschuss zu reduzieren.

Acknowledgment

This thesis was written during my work as a research assistant at the Institute of Energy and Process Systems Engineering at the Technische Universität Braunschweig. Most of all, I want to thank Prof. Dr.-Ing. Ulrike Krewer for including me into her institute, her supervision and support of my scientific work, and the productive atmosphere at the institute. In addition, I want to acknowledge the work of my board of examiners consisting of Prof. Dr.-Ing. Ulrike Krewer, Prof. Dr.-Ing. Richard Hanke-Rauschenbach, and Prof. Dr.-Ing. Arno Kwade, to examine my work and thesis, and to enable a save oral exam even in the current most difficult times of the Covid-19 pandemic.

Further, I acknowledge the work of the Battery LabFactory Braunschweig, their electrode production and cooperation in research projects. Exemplary, I want to mention and thank project partner Dr. Tom Patrick Heins and my student assistant Christoph Hirsch who took the measurements I used in my thesis to validate my simulations.

I acknowledge support from the German Federal Ministry of Economic and Affairs of Energy through funding projects "Data-Mining in der Produktion von Lithium-Ionen Batteriezellen (DaLion)" (03ET6089) and "FesKaBat - Feststoff-Kathoden für zukünftige Hochenergie-Batterien" (03ET6092D). This funding enabled my research as well as publication of results and presentation to the scientific community.

Last but not least, I want to thank all those great colleagues at the institute as well as my friends and family for supporting me during my time at the institute in many different ways. This includes my brother who steadily helped me improve my scientific English and never missed to remind me that sometimes working overtime is less important than working out.

Concluding, I want to thank everybody who accompanied my time at the institute. I am going to keep this time in fond memories.

List of Symbols

List of Symbols: Latin Letters

a_s	volume specific surface area in $\mathrm{m^2\,m^{-3}}$
A	area in $\mathrm{m^2}$
c	concentration in $\mathrm{mol\,m^{-3}}$
$\bar{c}$	normalized concentration
C	capacitance in $\mathrm{F\,m^{-2}}$
d	layer thickness in m
D	diffusion coefficient in $\mathrm{m^2\,s^{-1}}$
$\mathcal{D}$	parameter space $\subset \mathbb{R}^n$
E	potential in V
f	area specific weight in $\mathrm{kg\,m^{-2}}$
F	residual
F	Faraday's constant in $\mathrm{A\,s\,mol^{-1}}$
i_0	exchange current density in $\mathrm{A\,m^{-2}}$
I	current in A
j	current density in $\mathrm{A\,m^{-2}}$
k	reaction rate constant in $\mathrm{s^{-1}}$
k_i	adjustable parameter
l	domain size in m
L	thickness of full cell in m
$\mathcal{L}$	conductivity matrix of resistor network in S
$\mathcal{M}$	geometric structure
$\mathscr{M}$	model structure
n	number
$\mathcal{N}$	structure containing number of neighbors
r	radial coordinate in m
R_p	particle size in m
Q	charge in C
R	ideal gas constant in $\mathrm{J\,K^{-1}\,mol^{-1}}$
s	state vector
S	numerical particle size
S_i	Sobol index
S_{T_i}	total Sobol index
t	time in s
t_p	transference number
T	temperature in K
U	voltage in V
x	first spatial coordinate in m
x_0	starting point
X_i	uncertain generic parameter
$\bar{X}_i$	mean value of a generic parameter
Δx	voxel edge length in m
y	second spatial coordinate in m
$\mathcal{y}$	model output
z	third spatial coordinate in m
z	number of exchange electrons
Z	scalar function value

List of Symbols: Greek Letters

α	symmetry coefficient
β	Bruggeman coefficient
δ_{el}	layer thickness in m
ϵ	error
ε	porosity
ε_s	solid volume fraction
η	overpotential in V
ν	scalar parameter (distance to mean value)
ξ	sample point $\in \mathbb{N}^3$
ξ	linear coordinate in SEI in m
ζ	nucleus
ϕ	potential in V
φ	voxel position
σ	variance
σ_s	solid phase conductivity in $\mathrm{S\,m^{-1}}$
θ	weighting factor
θ^*	parameter set
Θ_s	lattice vacancy
Θ	parameter vector
Ω	domain/set of voxel

List of Abbreviations

AM	active material
ASSB	all-solid state battery
CB	carbon black
CBM	carbon black-binder matrix
CC	constant current
CV	constant voltage
DEM	discrete element method
DL	double layer
ECA	electron conducting additive
ECM	equivalent circuit model
EIS	electrochemical impedance spectroscopy
ESM	empiric surrogate model
FOM	factual order model
kMC	kinetic Monte Carlo
LIB	lithium ion battery
MC	Monte Carlo
OCP	open circuit potential
OCV	open circuit voltage
P2D	pseudo two dimensional
PCE	polynomial chaos expansion
PDF	probability density function
PE	parameter estimation
PEM	point estimate method
PEO	polyethylene oxide
SA	sensitivity analysis
SE	solid electrolyte
SEI	solid-electrolyte interface
SOC	state of charge
SPM	single particle model
UP	uncertainty propagation
UQ	uncertainty quantification

Chapter 1

Introduction

In this thesis, the battery cell production of lithium-based batteries is investigated. The dissertation addresses physics-based modeling in the context of electrode production. Electrochemical modeling is combined with three-dimensional micro structure modeling to quantify the influence of different aspects of electrode production on the cell performance.

1.1 Motivation

Nowadays, electricity generation from renewable sources and development of large-scale energy storage is growing rapidly. They are driven by environmental and ecological phenomena like an increased frequency of extreme weather conditions and air pollution due to human-induced emissions effecting billions of people. To reduce emissions, e.g. of CO_2, and thus global warming and air pollution, use of fossil fuels has to be substituted to a large extent. Further, increase of electricity production from time-variant wind and solar plants and the breakthrough of electric vehicles require storage of electrical energy. For applications with high requirements regarding energy and power density, the lithium-ion battery (LIB) has outperformed most other electrochemical storage technologies for about two decades now [1, 2].

For further increase of energy density, lithium foil anodes are aimed for as they have a high specific capacity and a low potential [3]. But due to safety and degradation issues of lithium foil anodes caused by dendrite growth, graphite intercalation anodes are commonly used. Today, different solid-state electrolytes are investigated as a possible candidate to reduce dendrite growth and enable the breakthrough of lithium foil anodes. With $600\,\mathrm{W\,h\,L^{-1}}$ and even more, all-solid-state batteries are believed to be a possible intermediate step to batteries with even higher theoretical energy densities [4]. Next-generation candidates are e.g. Li-air or Li-sulfur, which both are still subject of fundamental research but would allow energy densities beyond classical lithium-based intercalation batteries [3].

Current lithium-ion batteries with liquid electrolytes and graphite anodes can reach energy densities of up to $400\,\mathrm{W\,h\,L^{-1}}$ [4]. Till next-generation batteries potentially enable a steep increase of energy density, a steady enhancement of battery components and an optimization of battery production can increase the energy density and reduce waste and rejection rates and thereby contribute to reducing human-induced effects on the environment. Electrochemical battery modeling and process modeling can play an important role for optimization and understanding of cell performance and production. Beside mathematical optimization of cell design and production steps to reduce battery costs [5], an uncertainty quantification can reveal sensitive parameters and help to derive knowledge-based process bounds to reduce waste, rejection rates and cost. This application of models, electrochemical ones and others, in the production of lithium-based batteries lead to increased requirements regarding model accuracy. For instance, prediction of the calendering influence requires novel models to widen the range of validity of electrochemical models [6].

1.2 Scientific Question

Based on the global and societal importance of battery production and the development of modeling, the following hypothesis is stated: Mathematical modeling enables knowledge-based design of the battery cell production. This hypothesis leads to the need to assess the following question: Which model complexity is required to enable model-based process and design optimization, and which further requirements exist for such electrochemical models. To address this scientific question, applications and limitations of modeling in the context of production are investigated in this dissertation.

1.3 Outline

The outline is as follows: First, fundamentals of lithium-based batteries with liquid and solid electrolytes are reviewed. Then, modeling related to battery production is discussed. In Section 2.4, fundamentals of parameter estimation and identifiability are given. In Chapters 3 to 6, applications of models in production are addressed. In Chapter 3, parameters of the classical Doyle-Newman model are estimated and uniqueness of the derived parameter set is analyzed. In Chapter 4, uncertainty quantification of cell parameters related to electrode production is carried out to identify the relevant parameters and processes in cell production. In Chapters 5 and 6, two productions steps, mixing and calendering, are investigated simulation-based to highlight the potential of the application of models in the context of production. Therefor, a micro structure model is introduced which is used to enhance the structure-model parameter relations of the electrochemical model. The results of this work support the introduction of models to the production of LIBs and help to reduce cost, waste and rejection rates of production, while further increasing cell performance.

Chapter 2

Fundamentals

In this chapter, fundamentals of lithium-based batteries are introduced. This includes liquid electrolyte and solid-state electrolyte cells. In addition, cell production and modeling approaches are briefly reviewed.

2.1 Lithium-based Batteries

2.1.1 Cell Chemistry and Active Materials

Generally, a battery cell consists of anode, separator, electrolyte, and cathode. This structure is illustrated in Figure 2.1. The separator avoids an electrical short circuit

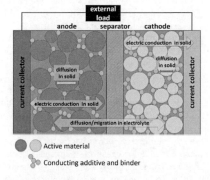

Figure 2.1: Schematic structure of a battery containing anode, separator and cathode. Large grey circles are active material, yellow circles represent the conducting additive-binder-domain and the pore space between the different particles is filled with electrolyte.

between anode and cathode, while the electrolyte enables ion transport between the electrodes. A conducting additive-binder domain ensures high electrical conductivity and mechanical stability of the electrodes. The current collectors connect active material and external electric circuit. Further, it provides mechanical stability to the electrode. Commonly, current collectors are made of copper at the anode side and aluminum at the cathode side. Different materials are used due to the different potentials at both electrodes and the electrochemical stability of aluminum and copper at the respective potentials.

The governing processes (see arrows in Figure 2.1) are solid state diffusion of lithium in the active material, diffusion and migration of ions in the electrolyte and electric conduction in the solid between the point of intercalation reaction and current collector.

For cells containing two intercalation electrodes, the main reactions at the particle surface can be summarized as

$$LiC_6 \leftrightharpoons Li_{1-x}C_6 + xe^- + xLi^+ \tag{2.1}$$

for the anode,

$$LiM \leftrightharpoons Li_{1-x}M + xe^- + xLi^+ \tag{2.2}$$

for the cathode, and

$$LiM + C_6 \leftrightharpoons Li_{1-x}M + Li_xC_6 \tag{2.3}$$

for the full cell. In Reaction 2.2, LiM denotes a lithium metal oxide or phosphate like lithium cobalt oxide ($LiCoO_2$, or LCO), mixed nickel-manganese-cobalt dioxide ($LiNi_{1-y-z}Mn_yCo_zO_2$, or NMC) or lithium iron phosphate ($LiFePO_4$, or LFP).

The most important properties of a battery are capacity, i.e. charge, and cell voltage, i.e. potential difference between anode and cathode. The capacity is limited by the utilizable concentration difference in the intercalation materials, and cell voltage by the potential difference between anode and cathode materials. Thus, an anode with a low potential and a cathode with a high potential are aimed for. Further important properties of batteries are a high discharge capacity at high current densities, low aging rates, thermal stability, and safety. Due to different weighting of those requirements, different active materials are used today [2].

A comprehensive review on cathode materials was published by Whittingham [7]. Therein, cell voltage, capacity, lattice structure, and volume changes due to intercalation are addressed. More recent reviews focussed on current trends in optimization of electrode materials [8], e.g. coating of LCO with magnesium and phosporus to enhance battery cycle life [8] and coating of NMC with SiO_2 or $Ti(OH)_4$ [9].

Today's development on cell chemistry and active material shows some general trends. At the anode, the silicon content in hybrid graphite-silicon electrodes is increased steadily to increase electrode capacity [10, 11, 12]. At the cathode, the cobalt content in mixed oxides like NMC is reduced due to increasing raw material cost of cobalt. Further coating, doping and compositing of active materials is applied to improve electric conductivity, diffusivity, and capacity [8]. And eventually, as safety is gaining more and more interest, liquid electrolytes are tried to be replaced by solid-state electrolytes.

2.1.2 Liquid and Solid Electrolytes

The objective of the electrolyte is to enable ion exchange between anode and cathode. Therefor, liquid electrolyte of LIBs consists of lithium salt, e.g. $LiPF_6$, and organic solvents like ethylene carbonate or dimethyl carbonate. Solid-state electrolytes are mainly of these types: polymers [13], oxides and sulfides [14]. Polymers are used in combination with lithium salt, while oxides and sulfides are ion conductors enabling lithium transport through their structure. A review about different ion transport mechanisms has been published by Aziz et al. [15].

Currently, there are only very few commercial all-solid-state batteries. They are commonly thin film batteries, wherein the cathode is a thin, non-porous electrode. This allows fast utilization of the active material but limits the volume specific energy density, due to a high volume fraction of inactive components like current collectors. The breakthrough of solid state electrolytes is retained by practical issues, like low cycle life due to mechanical degradation of the cell causing a rapid decrease of the electrochemical active interface area between solid electrolyte and active material [16]. While there are some materials with high conductivities at high temperatures [17], solid electrolytes, used commercially today, have low conductivities at room temperature. Research and development on solid-state electrolytes continuous as they e.g. could allow separator thicknesses of very few micro meters. This could increase the energy density at cell level and the thermal stability of solid electrolytes would increase the battery safety significantly compared to liquid electrolytes.

Beside issues in production and limited cycle life, further processes and effects have to be considered in design and modeling of all-solid-state batteries compared to cells with liquid electrolytes. While liquid electrolytes are commonly assumed to be an ideal solution of a salt and solvent, solid electrolytes consist of particles or grains, and dissolving the salt is often only possible at high temperatures. The grain boundaries lead to more complex transport mechanisms in the electrolyte [15]. While a grain boundary is an obstruction orthogonal to its orientation, the grain boundary could be an additional parallel transport path. In literature, there are first attempts to model those processes [18]. Modeling could enable the optimization of hybrid electrolytes which consist of a solid polymer electrolyte and an oxide or sulfide electrolyte [19, 20]. This concept could combine mechanical flexibility, provided by the polymer, with long cycle life and high conductivity provided by the oxide or sulfide. Yet, the application of hybrid electrolytes is limited by the interface resistance between both electrolyte phases. Further, the crystallinity of the polymer electrolytes has to be considered. In general, the more amorphous the electrolyte, the higher the conductivity. For instance, polyethylene oxide (PEO) is an insulator in its crystalline phase but becomes conductive in its amorphous state at high temperatures.

2.1.3 Battery Aging and Solid Electrolyte Interface

The aging behavior of a lithium ion battery is an essential performance criterion. As aging is beyond the scope of this dissertation, in the following the different aging mechanism are only briefly introduced. But the so-called solid-electrolyte-interface (SEI) will be explained in detail as it is relevant for investigation and assessment of pristine cells as well.

Commonly, aging is classified in calendaric aging, cycle induced aging and lithium plating. Calendaric aging summarizes electrochemical degradation during cell storage. It leads to formation of the SEI at the anode surface as the solvent of the electrolyte is reduced due to the low electrode potential [21]. This reaction leads to a capacity loss due to lithium bond in the SEI as well as to an increased cell resistance due to the SEI growth. The potential drop through this layer contributes to the inner cell resistance and is in constant current C-rate tests not distinguishable from the resistance contribution of other cell components.

Cycle induced aging summarizes aging during cell operation which contains different aging processes. It affects the active material at both electrodes and beside electrochemical degradation also leads to a mechanical degradation due to repeatedly expansion and shrinkage of active material due to repeatedly intercalation and deintercalation of lithium ions into the active material lattice structure [22].

Lastly, lithium plating is an aging process which is commonly related to fast charging or charging at low temperatures [23]. Both operation conditions allow the anode potential to drop to $0\,\mathrm{V}$ vs. $\mathrm{Li/Li^+}$. Then, deposition of metallic lithium become electrochemically more favorable compared to intercalation into graphite with its potential slightly above $0\,\mathrm{V}$ vs. $\mathrm{Li/Li^+}$.

For further details, the reader is referred to the various articles reviewing aging processes. E.g. Vetter et al. provided a comprehensive overview of aging phenomena [21]. More recently, electrochemical effects on the cathode's side are gaining interest as discussed in Ref. [24].

The aging processes summarized above occur at time scales of month to years, whereas lithium plating is commonly the most detrimental process if it is triggered due to respective operational conditions. However, the SEI is already formed during formation at the end of battery production and is only further growing with aging. Thus, it also affects investigation of pristine cells. Its effect on cell performance can be explained in the impedance spectra, as shown schematically in Fig. 2.2. At the left side, the distance

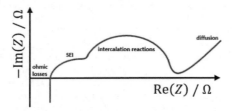

Figure 2.2: Schematic impedance spectrum of a lithium ion cell with assignment to different processes.

between ordinate and the intercept of the curve and the abscissa is the ohmic drop of the cell due to the resistance of solid electrodes and electrolyte[1]. The left semi-circle, at high frequencies, is related to electrochemical interface processes of the SEI, forming a double layer. At lower frequencies one or two semi-circles (one in Fig. 2.2) can be observed related to the main intercalation reactions of the two electrodes. Often, the reactions have similar time constants and thus their semi-circles are overlapping indistinguishably. For SEI and electrode reactions, the size of the semi-circle can be related to the reaction rate constant, namely exchange current density i_0 in the Butler-Volmer equation. Ideally, an electrochemical reaction would form a perfect semi-circle but due to irregularities in e.g. the particle surface the semi-circle becomes depressed as there is a variety of slightly different time constants. Last at low frequencies, the diffusion arc shows increase of resistance due to an increased polarization of the active material.

From this subsection, two things shall be kept in mind: First, for constant-current C-rate tests, the SEI has the same influence on the discharge performance as conductivity

---

[1]Everything below the abscissa is related to the inductance of testing equipment like wires and is thus of minor interest to the analysis of the cell itself.

of solid and electrolyte. Second, in contrast to C-rate tests, impedance spectra allow separation of ideal-ohmic resistances, SEI and intercalation reaction. This ability is of significant relevance for parameterization of electrochemical models. E.g., see Chapter 3 in this thesis.

2.1.4 Applied Reference Cell System

In the following, the electrochemical reference system considered throughout this dissertation is introduced.

The electrochemical system contains commercially available state-of-the-art materials. At the anode, surface-modified graphite (SMG from Hitachi Corp.) is used and at the cathode $Li(Ni_{1/3}Co_{1/3}Mn_{1/3})O_2$ (NCM of BASF Corp.). Additionally, Carbon Black (CB) and graphite (G) are added to both electrodes as conducting additives. The respective recipes are listed in Table 2.1. Electrodes with a diameter of 18 mm were manually punched out of sheets and assembled in a three electrode setup with a lithium metal ring reference (PAT-Cell, EL-CELL GmbH).[2] Electrode sheets were manufacture

Table 2.1: Recipes for cathode and anode suspension. Listed are the amounts of solid contents in weight-%.

	PVDF	CB	G	NMC, respectively SMG
Anode	5	2	2	91
Cathode	4	4	2	90

by the Battery LabFactory Braunschweig on a pilot-scale production line as described by Hoffmann et al. [25] and Laue et al. [26].

Layer thickness d_{el} and area specific weight f_{el} are measured, the specific capacity is derived from the open circuit potential (OCP) measurement, and mean porosity is mathematically derived from d_{el}, f_{el}, the composition of the electrode, and the densities of its components.

The electrolyte consists of EC, EMC and DMC with a ratio of 1:1:1 and traces of VC and CHB. For each cell 102.7 µL are used. All experiments were conduced at 25 °C in a temperature chamber. Formation of EL-Cells was conducted at 0.1C for two consecutive cycles. C-rate tests were conducted at 0.2C, 0.5C, 1C, 2C, and 3C. Discharge curves were recorded (Maccor 4000) between 4.2 V and 2.9 V. Electrochemical impedance spectroscopy was conducted in potentiostatic mode (VMP3, Bio Logic, France). For details it is referred to Heins et al. [27].

For OCP measurements, EL-cells were discharged incrementally from 4.2 V to 2.9 V in steps of 0.05 V and relaxation kept till $dU/dt \leq 0.2\,\mathrm{mV\,h^{-1}}$. The OCP in dependence on the normalized concentration $\tilde{c} = c/c_{\max}$ of the exemplary cell is described by

$$E_{\mathrm{OCP,a}}(\tilde{c}) = k_1 + k_2\tilde{c} + k_3\tilde{c}^{0.5} + k_4\tilde{c}^{1.5} + k_5\exp(k_6(k_7 - \tilde{c})) + k_8\exp(k_9(k_{10} - \tilde{c}))$$
$$+ k_{11}\exp(k_{12}(k_{13} - \tilde{c})) + k_{14}(k_{15} + \tilde{c})^{-1} \quad (2.4)$$

---

[2]For the electrochemical reference system, cell assemble was done by Georg Lenze and electrochemical experiments were executed by Tom Patrick Heins. The author gratefully acknowledges their technical support in the laboratory.

Table 2.2: Coefficients of the empirical half cell potential functions.

Coefficient	Anode	Cathode
k_1	8.0391 V	3.1058 V
k_2	5.0822 V	−9.5098 V
k_3	−12.561 V	10.372 V
k_4	0.4484 V	−5.0932 V
k_5	−0.0962 V	1.9426 V
k_6	15.001	−1.4960 V
k_7	0.1684	5.9019 V
k_8	−0.4599 V	$−1.3812 \times 10^{-5}$
k_9	2.3166	415.09
k_{10}	0.5856	−1.5383 V
k_{11}	−0.9575 V	-
k_{12}	2.4033	-
k_{13}	0.5124	-
k_{14}	−0.0114 V	-
k_{15}	0.03173	-

for the anode and by

$$E_{\text{OCP,c}}(\tilde{c}) = k_1\tilde{c}^6 + k_2\tilde{c}^5 + k_3\tilde{c}^4 + k_4\tilde{c}^3 + k_5\tilde{c}^2 + k_6\tilde{c} + k_7\exp\left(k_8\tilde{c}^{k_9}\right) + k_{10} \quad (2.5)$$

for the cathode, respectively. The respective coefficients are listed in Table 2.2. Equations, similar to those of Smith and Wang [28], are used and the coefficients are determined by a least square fit on the experimental data.

2.2 Processes for Large-Scale Battery Production

Lithium ion battery cell production can be separated in three parts: electrode production, cell assembling, and formation. In this section, relevant process steps are intro-

Figure 2.3: Flow chart of lithium ion cell production

duced. The individual process steps are pre-treatment/mixing, coating, drying, calendering, sheet separation, cell assembly, and formation. See Figure 2.3 for illustration.

2.2.1 Electrode Production

Electrode production of LIBs starts with pretreatment of the materials, e.g. deagglomeration of conducting additives and mixing of active material, binder, conducting additives like Carbon Black, and solvent. Length and intensity can have significant influence on electrode structure and cell performance [29, 30]. For all-solid-state electrodes, the production is slightly different since the electrolyte is mixed as well. This leads to more complex mixing routines as introduced in Ref. [31]. To optimize this process for ASSBs, a micro structure model is introduced in Chapter 5 of this work.

After mixing, the slurry is coated on the current collector, and the coating is dried. During the drying process, the solvent of the mixing process is removed from the electrodes. Again, the process parameters of drying have significant influence on the electrode structure and cell performance. Too fast drying can cause binder gathering at the electrode surface, increasing cell resistance and decreasing the mechanical stability of the electrode [32, 33].

Calendering is the final step of electrode production.[3] Therein, the electrodes are compressed between two rolls. This process has significant influence on the electrode structure and resistance. It provides a trade-off between ionic and electric conductivity [35]. For graphite anodes, this was shown e.g. by Shim and Striebel [36] and the negative effect of a too strongly compressed anode by Yang and Joo [37]. To understand the complex processes on micro scale in-depth, research focused on different aspects. Beside the decrease of the electrode resistance due to the Carbon Black network, the contact resistance between active material and current collector decreases significantly [38, 39]. The influence of calendering on pore size distribution and deformation of particles was investigated by Haselrieder et al. [35]. Schmidt et al. investigated highly compressed NMC electrodes and found an optimal high current performance at moderate porosities and a maximal energy density at low porosities [40]. A too low compression can be detrimental compared to no calendering, as binder or Carbon Black bridges between particles can be fractured [35, 41]. In addition, the calendering rate influences the wetting rate of the electrode which is important for cell production planing and utilization of the active material. Experimental results of Sheng et al. showed an optimal wetting at moderate calendering [42].

The effect of this process step is investigated in detail in this work. In Chapter 6, a micro structure model is coupled with a P2D model to quantify the calendering influence on cell performance.

2.2.2 Cutting and Cell Assembling

In large-scale battery production, coating, drying, and calendering steps are coil-to-coil processes. After calendering, electrode sheets are separated. A punching process is state-of-the-art while laser beam cutting is gaining interest as well [43]. Sheet separation is dependent on the desired cell format: cylindric, prismatic or stacked pouch cells. Separated sheets are then stacked or rolled in dependence on the cell format. Deviations in this process can lead to deviations in the utilized electrode area and can significantly affect the aging behavior of the cell [44].

The different cell formats have different benefits. Cylindric cells are widely applied in laptops and in some electric cars. They are easy to manufacture and have provided the highest energy densities for decades [2]. However, due to volume changes of the active materials, size and capacity of this cell type is limited. Especially in electric vehicles, prismatic cells are applied as well. This cell design allows additional safety features in the housing. The current trend goes to stacked pouch cells as they allow high capacities and high volume fractions of active materials [2]. Today, they are widely applied in

---

[3]The following paragraph about the calendering step was first published in a journal article which was written in the scope of this dissertation. See Ref. [34].

portable applications such as mobile phones.

2.2.3 Formation

After electrode production and cell assemble further non-mechanical process steps are following: wetting, self-discharge test[4], and formation. For wetting, the cell is stored to allow the electrolyte to enter the pores of the electrodes. Increased temperature can be applied to accelerate this process due to reduced viscosity of the electrolyte. For self-discharge test, the cell is stored in an electrically controlled state without actively charging or discharging the cell. Then, a high self-discharge over few days would reveal cells with internal faults, e.g. small internal short circuits, which allows to reject the cells before conducting long-term tests or selling the cells.

Cell production ends with formation. This is related to the first charging and discharge cycles at low C-rates. At first charging, the anode consists of pristine graphite. Its surface is electrochemically active with the electrolyte. Thus, a electrochemical side reaction is triggered at the anode which forms a solid layer at the particle surface, named solid electrolyte interface. Also see Section 2.1.3 for details about the SEI. It protects the anode against further side reactions, while being permeable for lithium ions. As the electrolyte of different cells, e.g. from different manufacturer, can contains different substances and additives, different reactions can occur leading to different SEI properties. Due to this interactions and the chance to optimize the SEI properties through design of electrolyte and electrical formation protocol, there is ongoing research to understand and optimize formation. For instance, Heimes et al. varied temperature and external mechanical load. The latter was shown to be able to reduce the required formation time [45]. An et al. designed a formation protocol with small amplitude cycles at high SOC. This approach was beneficial for cell performance as at high potentials different reactions occurs compared to low potentials [46]. Further, Röder et al. developed a kinetic Monte Carlo (kMC) multi scale model which showed the interaction of macroscopic and atomistic processes during SEI formation [47, 48, 49].

2.3 Modeling of Batteries and Battery Production

Decades of research led to a magnitude of models, which can be applied to understand and optimize the electrochemical performance, the battery production and its influence on cell performance. In this section, electrochemical models and some special models regarding the battery production are briefly reviewed.

2.3.1 Electrochemical Models

Electrochemical models can be classified by their discretization complexity, starting with non-discretized or 0D models, to pseudo-2D or P2D models, ending with 3D models and coupled multi-scale models. In works of Ramadesigan et al. [50] and Franco [51], models and their challenges were reviewed using similar classification.

A classification by model complexity and the related computational cost is illustrated qualitatively in Fig. 2.4.

---

[4]In literature, this process step is also denoted aging. However, to avoid inconsistency with cell aging, e.g. capacity fade, during cell operation or long time storage, the term self-discharge test is used in this work.

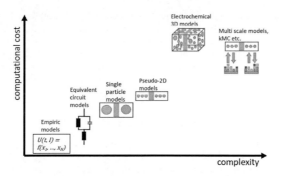

Figure 2.4: Computational cost versus model complexity

Sub-categories of non-discretized models are lumped electrochemical models, empiric models and equivalent circuit models. Equivalent circuit models (ECM) consider a set of electrical components like resistors, capacitors or constant phase elements to simulate the short-term voltage response of a battery to a current input or the other way around. Those models are commonly applied for control purposes e.g. to optimize fast charging protocols [52, 53].

The most common models nowadays are single particle models (SPM) or pseudo-2D models based on the work of Doyle, Fuller and Newman [54, 55]. Those models consider discretized charge and mass transport in one linear coordinate for the electrolyte and one radial coordinate in the active material particle. The SPM therein considers one representative particle per electrode and the P2D model one particle per volume element of the discretized electrolyte. Governing equations of the P2D model and the single particle model are introduced in Sections 2.3.2 and 2.3.3, respectively.

Doyle-Newman-type models have been applied frequently and have been adjusted to the respective investigation objective. For instance, the solid phase diffusion equations were extended to consider diffusion induced mechanical stresses and the influence of stress on diffusion [56]. Ramadass et al. included a first aging model, first for calender aging [57], then for cycle aging [58]. Further, more and more aging mechanisms were included: mechanical failure leading to increased active surface area [59] and lithium plating [60]. In dependence on the active material, phase changes due to the lithium concentration were modeled, as for instance silicon particles pass through discrete intercalation steps leading to interfaces with additional mechanical stresses [22, 61]. Also, the influence of concentration dependent diffusion coefficients [62] and particle size distribution [63] on the discharge performance were addressed.

Electrochemical full-3D models became more common as computational power increased significantly in the last decade. For instance, they enable to study the effect of spatial current distributions on thermal effects [64]. Electrochemical full-3D models may consider micro structures based on generation of artificial structures or reconstruction of e.g. micro computed tomography scan images or scanning electron microscope images. Those models are able to address the influence of local effects like pore size distribution [65].

Kinetic Monte Carlo models are listed beyond full-3D regarding model complexity as they often consider a 3D kMC nanometer scale model which is coupled to a macroscopic cell model [66]. In kMC, stochastic effects and random movement of the reactants at the active material surface are modeled. These models are applied to investigate aging [67] or the solid electrolyte interphase generation during the formation steps at the end of production [66]. The electrochemical models reviewed above simulate performance of a cell or a cell part after the cell was completely manufactured.

2.3.2 Governing Equations of the P2D-Model

A general review about electrochemical modeling of batteries is given in Section 2.3.1. In the following, governing equations of the electrochemical model applied throughout this dissertation are introduced in detail. Generally, discretization is done in x-direction from anode to cathode and in r-direction from the particle center to the particle surface. As there is no interaction between the elements of the second dimension (particle radius), this is termed pseudo-two-dimensional. See Fig. 2.5 for illustration. The outlined domain boundaries are explaining in the further.

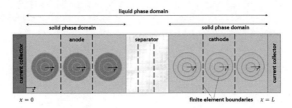

Figure 2.5: Schematic structure of P2D model and the mathematical domains of solid and liquid phase.

The second and third dimension of the liquid phase is neglected as there is no significant mass transport in y- or z-direction

$$\frac{\partial c_e}{\partial y} = \frac{\partial c_e}{\partial z} = 0 \tag{2.6}$$

as the cell area A, in y- and z-directions, is commonly huge compared to the anode's and cathode's thickness $\delta_{\text{el,a}}$ and $\delta_{\text{el,c}}$ in x-direction, and as homogeneous electrode properties are assumed [54]. The solid electrolyte interface is not considered explicitly in this model.

A dual intercalation cell is implemented considering the reaction

$$\text{Li}^+ + e^- + \Theta_s \rightleftharpoons \text{Li} - \Theta_s \tag{2.7}$$

at both electrodes wherein Θ_s is a free space lattice in the intercalation material (solid) and $\text{Li} - \Theta_s$ is an intercalated lithium atom [68]. The reaction kinetics are described using a Butler-Volmer-type expression

$$j^{\text{Li}} = a_s i_0 \left(\exp \left(\alpha \frac{\eta \text{F}}{\text{R}T} \right) - \exp \left(-(1 - \alpha) \frac{\eta \text{F}}{\text{R}T} \right) \right) \tag{2.8}$$

wherein the exchange current density i_0 is concentration dependent:

$$i_0 = kF c_e^\alpha \left(c_{\max} - c_s\right)^\alpha c_s^{1-\alpha}. \tag{2.9}$$

In Eq. 2.8, α is the symmetry coefficient of the electrochemical reaction, η is the reaction overpotential, F is Faraday's constant, R is the ideal gas constant, T is temperature and a_s is the volume-specific active surface area derived from particle size R_p and active material volume fraction ε_s:

$$a_s = \frac{3 \cdot \varepsilon_s}{R_p}. \tag{2.10}$$

This equation is based on the assumption of mono-disperse spherical particles without any particle-to-particle contacts. Thus, it validity range is limited to $\varepsilon \in [0, 0.74]$, wherein the upper bound is related to close-packing of equal spheres. The reaction fluxes $j^{Li}(x)$ occurring at the particle surface provide the boundary condition at $r = R_p$ for the mass transport processes in the solid particle

$$\frac{\partial c_s}{\partial r} = \frac{-j^{Li}}{z \cdot F \cdot a_s \cdot D_s}, \; r = R_p \tag{2.11}$$

wherein c_s is the lithium concentration inside the active material, z is the number of transferred electrons in Reaction 2.7 and D_s is the diffusion coefficient of lithium inside the active material. In the center of the rotationally symmetrical particle the boundary condition is zero-flux:

$$\frac{\partial c_s}{\partial r} = 0, \; r = 0. \tag{2.12}$$

The diffusion inside the spherical solid domain is modeled by Fick's second law

$$\frac{\partial c_s}{\partial t} = \frac{1}{r^2} \frac{\partial}{\partial r} \left(D_s r^2 \frac{\partial c_s}{\partial r} \right). \tag{2.13}$$

Mass transport in the electrolyte considers migration and the electrochemical reactions as a source term in addition to diffusion

$$\varepsilon \frac{\partial c_e}{\partial t} = \frac{\partial}{\partial x} \left(D_{e,\text{eff}} \frac{\partial c_e}{\partial x} \right) + (1 - t_p) \frac{j^{Li}(x)}{F} \tag{2.14}$$

wherein c_e is the lithium ion concentration in the electrolyte and $D_{e,\text{eff}}$ is the effective diffusion coefficient of lithium ions in the electrolyte, which is derived from the bulk diffusion coefficient $D_{e,0}$, porosity and Bruggeman coefficient β:

$$D_{e,\text{eff}} = D_{e,0} \cdot \varepsilon^\beta. \tag{2.15}$$

Boundaries of the liquid domain are the two current collectors at $x = 0$ and $x = L$ (see Fig. 2.5), whereat L is the combined thickness of both electrodes and the separator. The zero-flux boundary conditions are as follows:

$$\frac{\partial c_e}{\partial x} = 0, \; x = \{0, \; L\}. \tag{2.16}$$

The liquid phase potential ϕ_e is calculated through

$$j^{\mathrm{tot}} = -\frac{\partial}{\partial x}\left(\sigma_{\mathrm{e,eff}}\frac{\partial \phi_e}{\partial x}\right) - 2\frac{RT}{F}\left(t_{\mathrm{p}} - 1\right)\sigma_{\mathrm{e,eff}}\frac{\partial \ln c_e}{\partial x} \tag{2.17}$$

wherein the total flux j^{tot} is the sum of the reaction flux j^{Li} and the double layer flux j^{DL}

$$j^{\mathrm{tot}} = j^{\mathrm{Li}} + j^{\mathrm{DL}} \tag{2.18}$$

with

$$j^{\mathrm{DL}} = a_s C_{DL}\frac{\partial\left(\phi_s - \phi_e\right)}{\partial t}. \tag{2.19}$$

The boundaries of the liquid phase domain are the current collectors. Both boundary conditions are zero flux:

$$\frac{\partial \phi_e}{\partial x} = 0, \ x = \{0, \ L\}. \tag{2.20}$$

Solid phase potential ϕ_s is governed by Ohm's law

$$j^{\mathrm{tot}} = \frac{\partial}{\partial x}\left(\sigma_{\mathrm{s,eff}}\frac{\partial \phi_s}{\partial x}\right) \tag{2.21}$$

wherein $\sigma_{\mathrm{s,eff}}$ is the effective solid phase conductivity derived from bulk conductivity $\sigma_{\mathrm{s,0}}$ and solid phase volume fraction ε_{s}:

$$\sigma_{\mathrm{s,eff}} = \sigma_{\mathrm{s,0}} \cdot \varepsilon_{\mathrm{s}}. \tag{2.22}$$

The solid phase domain is limited to the electrodes as the separator is neither conductive to electrons nor contains active material. Thus, there are four boundaries (two per electrode) of solid phase potential. At the current collector-to-active material interfaces (at $x = 0$ and $x = L$), the potential gradient is governed by the applied cell current

$$\frac{\partial \phi_s}{\partial x} = \frac{-I_{\mathrm{cell}}}{A_{\mathrm{cell}} \cdot \sigma_{\mathrm{s,eff}}}, \ x = \{0, \ L\} \tag{2.23}$$

and at the active material-to-separator interfaces there are zero-flux boundary conditions:

$$\frac{\partial \phi_s}{\partial x} = 0, x = \{\delta_{\mathrm{el,a}}, \ L - \delta_{\mathrm{el,c}}\}. \tag{2.24}$$

In conclusion, the model has a set of time-dependent state variables, i.e., solid and liquid phase concentrations c_s and c_e, respectively, solid and liquid phase potentials ϕ_s and ϕ_e, and the surface overpotential η for both electrodes, respectively. From the states of the solid phase potential, the cell voltage can be derived

$$U_{\mathrm{cell}} = \phi_{\mathrm{s}}(L) - \phi_{\mathrm{s}}(0) \tag{2.25}$$

In order to reduce the number of independent uncertain parameters, the Nernst-Einstein equation is considered

$$\sigma_e = \frac{F^2}{R} \cdot c_{\mathrm{Li}} \cdot (D^+ + D^-) \tag{2.26}$$

which describes the relation between liquid phase conductivity, σ_e, and cation's and anion's liquid phase diffusion coefficient, D^+ and D^-, respectively. This is a simplification compared to the concentrated solution theory of the classic Doyle-Newman Model [54]. However, it is of minor significance at the applied concentrations.

2.3.3 Governing Equations of the Single Particle Model

The single particle model is a computational more efficient alternative to the P2D model. In this work, a SPM is applied to simulate the electrochemical impedance spectroscopy (EIS). In Fig. 2.6, the structure of the single particle is illustrated. The main difference to the P2D model (see Fig. 2.5 for illustration) is the discritization of the liquid phase domain in only three finite elements: one for anode, separator and cathode, respectively. In consequence, there are only two particles which are discritized with $n > 1$ elements. In general, the same governing equation as introduced in Section 2.3.2 are applied. This simplification has a small influence on the model precision at low C-rate, where

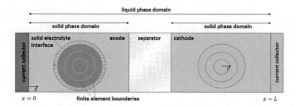

Figure 2.6: Schematic structure of single particle model and the mathematical domains of solid and liquid phase.

the polarization of the electrodes is small. This is especially valid for the simulated impedance spectra in this work.

Beside the reduced computational time, the applied SPM is especially feasible to model the impedance spectrum, as it models the SEI more in detail than the P2D model introduced above. An additional charge and mass transport through the SEI, as well as an additional adsorption step is considered [69, 70]. The additional adsorption kinetic is as follows

$$r_{\text{Li,ads}} = \Theta_{\text{v}} a_{\text{s,ads}} c_{\text{Li,e}} k_{\text{b}} \exp \left(-\alpha \Delta \phi \frac{\text{F}}{\text{R}T} \right)$$
$$-\Theta_{\text{Li}} a_{\text{s,ads}} k_{\text{f}} \exp \left(\alpha \Delta \phi \frac{\text{F}}{\text{R}T} \right)$$
(2.27)

wherein Θ_{v} is a vacant lattice space in the SEI and Θ_{Li} is a lithium-filled vacancy.

Further, the SEI is discretized regarding diffusion and solid potential. As the layer is thin, a linear coordinate is chosen instead of a radial coordinate as in the spherical particle:

$$\frac{\partial c_{\Theta_{\text{Li}}}}{\partial t} = \frac{\partial}{\partial \xi} \left(D_{\Theta_{\text{Li}},\text{SEI}} \frac{\partial c_{\Theta_{\text{Li}}}}{\partial \xi} \right),$$
(2.28)

$$j^{\text{Li, ads}} = \frac{\partial}{\partial \xi} \left(\sigma_{\text{SEI}} \frac{\partial \phi_{\text{SEI}}}{\partial \xi} \right).$$
(2.29)

In Eqs. 2.28 and 2.29, the linear coordinate ξ, orthogonal to the particle surface, is introduced.

2.3.4 Modeling the Calendering Influence[5]

In terms of calendering, there are two different modeling approaches. On the one hand, there are process models to describe the production process itself. On the other hand, there are electrochemical models which are used to quantify the influence of calendering on the electrochemical cell performance.

Process models can be categorized into empiric models which are strongly related to experimental data [71], and highly sophisticated physics-based models, e.g. based on the discrete element method (DEM). The physics-based models , e.g. simulate the particle movement during the calendering step to gain insight into processes as well as to determine the optimal calendering rate [72].

Electrochemical modeling is widely applied. However, only very few contributions are specifically assigned to calendering. Lenze et al. applied a P2D model to estimate the effective electrode properties at different calendering rates and revealed significant changes of electric conductivity and active surface area in dependence on the calendering rate [73]. Kenney et al. quantified the impact of deviations in the drying and calendering process, applying an electrochemical single particle model [74] and Smekens et al. simulated the calendering influence on a positive electrode [39].

Micro structures have been modeled frequently but mostly without focus on calendering. Some models are based on artificial structures [75, 76, 77, 78], some reconstruct structures from e.g. images of sccanning electron microscope of ion beam cut samples [79, 80, 81, 82, 83]. There are DEM micro structure models, e.g. see Ref. [84], FEM micro structure models, e.g. see Ref. [65], and full-3D electrochemical models [85, 86, 81, 65]. The later suffer from excessive computational costs even at domain sizes of very few particle sizes [78]. Beside stochastic approaches and reconstruction, Liu and Mukherjee used a kinetic Monte Carlo method to investigate the influence of solvent evaporation and interaction of solvent and nano particles on the conductive interfacial area [87]. Ngangdjong et al. introduced a multiscale approach of coarse grain molecule dynamic simulations coupled with a electrochemical full-3D model [88]. This approach allowed to predict the micro structure in dependence on the fabrication process and to simulate the electrochemical cell performance.

2.3.5 Simulation in Production

While the electrochemical models described above are commonly modeling the battery from particle to cell level, research on production led to a variety of model scales up to factory level. In this section, process models and process chain models are reviewed briefly. However, process models are less common than electrochemical models.

Process models have been introduced for different individual machines, process steps or rooms. Solvent evaporation during the drying process and the influence on the spatial binder distribution have been modeled by Liu and Mukherjee [33]. The calendering step is investigated with process models based on the discrete element method modeling the forces and movements of particles during the calendering process [72]. Those models allow to optimize the process to get a desired electrode structure. Knoche et al. proposed a process model for the electrolyte filling based on a graphical approach which allows to derive qualitative results and guidelines for production planing [89]. Ahmed et al.

---

[5]This subsection was first published by Laue et al. [34]

modeled the humidity control of a dry room focusing on energy and cost per battery pack [90].

On a larger scale, there are process chain models considering different process steps and the entire production building. Simulation models for manufacturing were reviewed by Negahban and Smith [91]. Process chain models have been used for management and planing purposes [92] and to quantify uncertainty propagation [93]. The latter combined an agent-based process chain model with an electrochemical battery model. Papadopoulos et al. reviewed approaches to model manufacturing systems with stochastic Markov models [94].

2.4 Fundamentals of Parameter Estimation and Identifiability[6]

While parameter estimation (PE) is commonly used to parameterize models, it also is a powerful tool to derive non-measurable parameters from cell performance. The later was introduced as simulation-based diagnosis [73]. Both applications require practical identifiability of the model parameters. In this chapter, parameter estimation approaches are briefly reviewed, and direct and sample-based approaches of proofing identifiability are summarized.

2.4.1 Parameter Estimation

Common estimated parameters Different types of parameters are estimated. The most common parameters are reaction rate constants, diffusion coefficients of active materials and electrolyte, and conductivity of solid and electrolyte. To model temperature dependencies commonly Arrhenius' law is applied, which has a pre-exponential coefficient and an activation energy which are estimated from temperature variations. Further, consideration of concentration dependency may increase the number of parameters. For instance, in Ref. [96], the state of charge (SOC) dependent, respectivly concentration dependent, open cell voltage (OCV) of both electrodes was tried to be estimated from the electrochemical experiments leading to a large number of parameters, while in Refs. [26] and [73], the OCV was parameterized prior to the parameter estimation of the common model parameters, leading to a small number of parameters.

For equivalent circuit models, resistances, capacitance, and OCV have to be estimated. Here, concentration dependencies are common.

Common amount of estimated parameters The number of estimated parameters varies significantly between different publications, as summarized in Table 2.3. This is partially related to different model complexities and types, and partially to the different available experimental data. Noticeable in Table 2.3 is the large number of 88 parameters in Ref. [96]. This number is related to the try to estimate both OCV curves simultaneously with the other parameters. However, they only could estimate half of their parameters as only full-cell experiments were available. Thus, they could solely estimate the OCV difference between the two electrodes. Further, they estimated geometric properties of the electrodes. As Table 2.3 shows, the most common case is a P2D model without temperature dependency in combination with C-rate tests. This commonly leads to three to 14 parameters. The different number of parameters are partially

---

[6]This section was first published in a jorunal article within the scope of this dissertation. See Ref. [95].

Table 2.3: Number of estimated parameters in different publications in order of publication. Used abbreviation are: double layer (DL), single particle model, pseudo-2D model, equivalent circuit model, constant voltage (CV), and constant current (CC).

number of parameters	year	Ref.	model type	experiment
88	2012	[96]	SPM incl. OCV	drive cycles
11	2013	[97]	reduced-order SPM	C-rate and pulse tests
18	2014	[98]	P2D incl. T-dependency	C-rate tests
4	2015	[99]	P2D	one charge/discharge
3	2017	[73]	P2D incl. DL	C-rate test
27	2018	[100]	P2D	C-rate tests
6	2018	[101]	ECM	C-rate test
14	2018	[102]	P2D	CC/CV charge/discharge
7	2019	[26]	P2D incl. DL	C-rate test

related to different parameterization approaches and research objectives, but also raise the equation about the maximum number of parameters which can be estimated in one model. To assess the capabilities of parameter estimation and to identify its limitation, different approaches of PE are reviewed in this section and in Section 2.4.2 fundamentals of identifiability are briefly summarized.

Preliminary work For physics-based full-order models, Ramadesigan et al. applied least-squares fitting to estimate parameter changes due to cell aging in combination with an uncertainty quantification to calculate the confidence intervals of the estimated parameters [50]. Vazquez-Arenas et al. coupled the parameter estimation with a sensitivity analysis to assess the estimation accuracy [98]. Lenze et al. estimated parameters of an electrochemical model manually adjusting parameters [73]. Chun and Han introduced a cascaded improved harmony search to estimate the parameters of the P2D model, wherein a partially random-based optimization approach was conducted repeatedly [102].

In the engineering and control community, there exists a variety of reduced order models or equivalent circuit models [103, 104, 105, 106] and parameter estimation is conducted frequently [103]. Focus of these models is commonly estimation of battery states like SOC, state of power, or state of health [106, 107]. Dvorak et al. introduced a parameter estimation method for an ECM containing 2 RC-elements[7]. This method applies step-wise discharge at constant current with repetitions at different C-rates and temperatures [101]. For ECMs as well, it was shown that parameter estimation with a Kalman filter should facilitate electrochemical tests similar to the actual application for optimal accordance of the model [108].

Beside the choice of the experimental electrical test procedure and optimization algorithms, in few publications the objective function was enhanced in terms of applying different objective functions to different parts of the experimental data [109, 110]. For instance, Li et al. applied three steps: high SOC and low current, low SOC and low cur-

---

[7]A RC-Element is a resistance in paralell to a capacitance.

rent, and eventually a dynamic driving cycle [110]. Jin et al. considered thermodynamic parameters and kinetic parameters independently for a P2D model as well as linear or logarithmic scaling for different parameters [100]. All these publications have a strong theoretical focus. In contrast, Ecker et al. used as many experimental analysis methods as possible to measure e.g. diffusion coefficients, and used electrochemical impedance spectroscopy in combination with equivalent circuit models to partially estimate the parameters of the P2D model [109, 111]. Schmalstieg et al. presented a method to parameterize a model of bought-in commercial cells [112]. This method combined EIS, galvanostatic intermittent titration technique, porosity measurement by mercury intrusion, and ECM fitting [112].

2.4.2 Identifiability of Model Parameters

Identifiability describes the property of a model, whether its parameters can be identified from its output. Bizeray et al. defined itdentifability as follows [113]:

Definition *Consider a model structure $\mathcal{M}$ with an output $\mathcal{Y}(s, \theta)$ parameterized by $\theta \in \mathcal{D} \subset \mathbb{R}^n$ where n denotes the number of parameters of the model. The identifiability equation for $\mathcal{M}$ is given by*

$$\mathcal{Y}(s, \theta) = \mathcal{Y}(s, \theta^*) \quad \text{for almost all } s, \tag{2.30}$$

wherein θ, $\theta^* \in \mathcal{D}$. The model structure $\mathcal{M}$ is said to be

- *globally identifiable if Eq. 2.30 has a unique solution in $\mathcal{D}$,*

- *locally identifiable if Eq. 2.30 has a finite number of solutions in $\mathcal{D}$,*

- *unidentifiable if Eq. 2.30 has an infinite number of solutions in $\mathcal{D}$.*

The three kinds of identifiability are illustrated in Fig. 2.7. A minimum represents a solution of Eq. 2.30. For $n = 1$, the unknown parameter is varied. In the left plot,

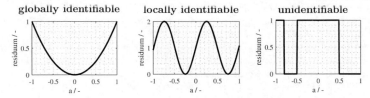

Figure 2.7: Types of identifiability for one parameter a. The y-axis represents the residuum between simulation and experiment. Thus, a minimum close to zero represents a parameter set leading to a high accordance between experiment and simulation.

global identifiability is given, as there is only one solution in $\mathcal{D}$. In the central plot, local identifiability is illustrated, as there is a finite number, two in this case, of solutions. Unidentifiability is illustrated in the right plot, as there is an infinite number of solutions for $a \in [-0.8, -0.5]$ and $a > 0.5$.

Further, the terms structural identifiability [113], practical identifiability [99] and uniqueness are introduced. Structural identifiability focuses on the model function itself and assesses whether identifability is given or not. In general, it refers to global identifiability. Practical identifiability also considers the sensitivity of $\mathcal{M}$ to θ. Practical identifiability requires additionally sensitivity and is related to local identifiability. Uniqueness of the estimated parameter set, respectively solution of Equation 2.30, yields global identifiability.

Table 2.4: Number of identified parameters in literature.

model	identified	Ref.
ECM	2 out of 2	[114]
SPM	half of 88	[96]
P2D	3 out of 8	[115]

Table 2.4 briefly summarizes the percentage of identifiable parameters in literature. See Section 2.4.1 in this work for a discussion about common estimated parameters and their amount. Sharma and Fathi [114] and Lin and Stefanopoulou [116] used the Fisher information matrix to assess local identifiability. From the Fisher information matrix the eigenvalues can be used to assess identifiability [117, 118], as well as a singular value decomposition can be carried out. Bizeray et. linearized a single particle model to investigate its structural identifiability [113]. This approach has the advantage to be able to assess global identifiability and hence uniqueness, but requires a closed formulation of the models output and linearization. Thus, it is not applicable to the commonly used P2D battery model which lacks the closed formulation of its output. Also, it is possible that linearization reduces the model's identifiability. In contrast, the Fisher information matrix could be applied for a P2D model, but computational cost would be enormous for a state-of-the-art desktop PC. For a SPM the Fisher information matrix was successfully applied by Pozzi et al. to derive optimal experiments for parameterization [119].

Barcellona et al. published a review on parameter identification techniques for lithium ion battery models [103]. Therein, they were classified in online methods, offline methods or analytically/numerical calculation methods. Online methods were commonly applied to ECMs for state estimation [103]. Tian et al. estimated parameters of an ECM from constant-current discharge data and assessed local identifiability analyzing the rank of the sensitivity matrix [120]. Pozna et al. showed that even a first order RC equivalent circuit battery model is not identifiable [121]. For fractional-order models (FOM) of lithium-ion batteries, parameter identifiability was assessed by Guo et al. and Li et al. [122, 110]. FOM is based on Laplace transformation and provides lumped parameters which have to be estimated.

2.5 Fundamentals of Uncertainty Quantification[8]

The manufacturing process of lithium-ion batteries contains many process parameters. Their influence on the product performance is often nonlinear and not well understood, especially quantitatively. However, to achieve a product that meets performance and

---

[8]This section was first published in a journal article which was written in the scope of this dissertation. See Ref. [26].

quality goals such as high energy density and low costs, the manufacturing process has to be optimized, and uncertainties of process and product parameters have to be dealt with [123]. Due to the crucial need for an advanced process understanding, the application of uncertainty quantification (UQ) methods to the manufacturing process of LIBs has gained interest [124, 125, 74, 126]. In extend to the common modeling approaches as reviewed in Ref. [127], UQ provides quantitative information about the interaction between different parameters and could enable adaption of production quality measures to the specific produced battery. As a model-based approach, UQ can partially substitute the frequently-used time- and cost-intensive experimental approaches.

2.5.1 Preliminary Work

In the context of lithium ion batteries, there are different approaches of uncertainty quantification or uncertainty propagation (UP). First, there are sample-based, or direct, methods and second, there are indirect methods which adapt the system. The sample-based methods contain e.g. Monte Carlo (MC) simulations wherein the system is evaluated with random parameter sets. For instance, Kenney et al. quantified the influence of process variations concerning layer thickness, electrode density and the amount of active material by evaluating a serial set of single particle models with randomly chosen parameter sets in each model [74]. Indirect methods contain Polynomial Chaos Expansion (PCE), see e.g. Ref. [128], wherein variations of the particle size were addressed, and methods of control theory as used in Ref. [129], wherein an analytic model of the battery's impedance response was applied considering variations of porosity, particle size and layer thickness. The first is limited to a low number of uncertain parameters and the later solely provides upper and lower bounds of the uncertain output [128].

Beside the usage of UQ to investigate the UP between product parameters and cell performance, UQ also can provide a sensitivity analysis (SA) for parameter estimation and parameterization of physics-based models [98, 50]. Especially for parameterization, the number of uncertain parameters is large compared to UP investigations. For instance, Schmidt et al. considered 33 uncertain parameters in a single model. They applied SA based on the Fisher-information matrix [130]. More recently, Lin et al. applied polynomial chaos expansion to a 3D multiphysics battery model to conduct a global sensitivity analysis [131].

In literature, different distribution shapes and widths are observed. E.g. experiments of Schuster et al. and Dubarry et al. showed deviations of quantities like weight, capacity and resistance variations from $\pm0.1\%$ to $\pm5.7\%$ considering partially normal and partially Weibull distributions [132, 133], while Vazquez-Arenas et al. assumed a normal distribution with variations of $\pm10\%$, see Ref. [98], and Kenney et al. investigated electrodes assuming variances between 1% and 5%. See Ref. [74]. Recently, Röder et al. investigated the impact of a particle size distribution and its shape considering a Weibull distribution [63].

As listed above, common uncertain parameters are layer thickness, porosity, and particle size. It is noteworthy that there are significant differences between distributions of macroscopic quantities like layer thickness and microscopic quantities like particle size. In general, four different kinds of parameter distributions have to be considered. The scale of those deviations is illustrated in Fig. 2.8. First, parameters may be distributed on particle level due to the processing before the electrode and cell production, e.g. particle size [134, 62, 63]. Also, its distribution corresponds to the used active material.

Second, there are local changes of macroscopic quantities like porosity due to small scale process deviations. Those occur at the scale of a single electrode sheet or below, termed sub-cell, and they are influenced by process steps like mixing [29, 135, 136], calendering [35, 136, 137, 138, 38, 37, 139] and drying [140]. Third, there are cell-to-cell deviations of macroscopic quantities like layer thickness [125, 74, 126, 132, 133]. Those are caused by process deviations, which occur at large scales compared to the size of the electrode of a whole cell. Fourth, there are lot-to-lot or batch deviations occurring between lots

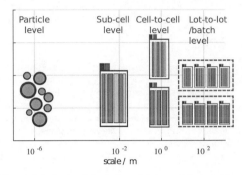

Figure 2.8: Scale of deviations from particle to lot level

or batches produced e.g. at different days [132, 129].

2.5.2 Point Estimate Method

The point estimate method (PEM) is one tool for an uncertainty quantification. It was introduced by Tyler [141] in 1953 and Rosenblueth [142] in 1975 and is frequently used, e.g. by Lin and Li [143], to reconstruct the probability density function (PDF) of an output variable Z of a non-linear function

$$Z = F(X_1, X_2, \cdots, X_N) \in \mathbb{R}^N \qquad (2.31)$$

based on a finite number of sampling points of the N variables $X_1, X_2, \cdots, X_N$. As illustrated in Fig. 2.9, different uncertain input variables with different distribution widths can be considered. While F can be a black box, differentiability of F is beneficial as a non-differential function could cause physically meaningless results [143].

In the following, an UQ framework is used, introduced as PEM5 in Ref. [144] and references therein. This method requires $2N^2 + 1$ sampling points, which is significantly less than the number of sample points a Monte Carlo approach would require normally. For instance, assuming three uncertain parameters X_1, X_2, X_3, which are normally distributed with $\bar{X}_i = 0, \forall i \in \{1, \cdots, 3\}$, there are 19 sample points $\xi_i, \forall i \in \{1, \cdots, 19\}$. The three generator functions $GF[\cdot]$ of the sample points are

$$GF[0] = \{(0, 0, 0)^{\mathrm{T}}\}, \qquad (2.32)$$

$$\begin{aligned} GF[\pm\nu] = \{&(\nu, 0, 0)^{\mathrm{T}}, \ (-\nu, 0, 0)^{\mathrm{T}}, \ (0, \nu, 0)^{\mathrm{T}}, \\ &(0, -\nu, 0)^{\mathrm{T}}, \ (0, 0, \nu)^{\mathrm{T}}, \ (0, 0, -\nu)^{\mathrm{T}}\}, \end{aligned} \qquad (2.33)$$

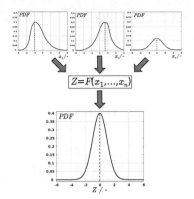

Figure 2.9: Flow chart of uncertainties for a generic function $Z = F(X_1, \ X_2, \ X_3) \in \mathbb{R}^3$

and

$$
\begin{aligned}
GF[\pm\nu, \pm\nu] = \{ &(\nu, \nu, 0)^{\mathrm{T}}, \ (-\nu, -\nu, 0)^{\mathrm{T}}, \ (\nu, -\nu, 0)^{\mathrm{T}}, \\
&(-\nu, \nu, 0)^{\mathrm{T}}, \ (\nu, 0, \nu)^{\mathrm{T}}, \ (-\nu, 0, -\nu)^{\mathrm{T}}, \\
&(\nu, 0, -\nu)^{\mathrm{T}}, \ (-\nu, 0, \nu)^{\mathrm{T}}, \ (0, \nu, \nu)^{\mathrm{T}}, \\
&(0, -\nu, -\nu)^{\mathrm{T}}, \ (0, \nu, -\nu)^{\mathrm{T}}, \ (0, -\nu, \nu)^{\mathrm{T}} \}
\end{aligned} \tag{2.34}
$$

wherein ν is a scalar parameter.

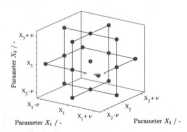

Figure 2.10: Position of the $2N^2 + 1$ sample points ξ_i for a generic function $Z = F(X_1, \ X_2, \ X_3) \in \mathbb{R}^3$

In general, the first sample point, ξ_1, would be $(\bar{X}_1, \bar{X}_2, \bar{X}_3)^{\mathrm{T}}$ and the second sample point, ξ_2, would be $(\bar{X}_1 + \nu, \bar{X}_2, \bar{X}_3)^{\mathrm{T}}$. But as $\bar{X}_i = 0, \forall i$, all $\bar{X}_i$ are spared in Eqs. 2.32 to 2.34. Fig. 2.10 illustrates the positions of these sample points for the 3D example. Introducing weights $w_j, \forall j \in \{0, \cdots 2\}$, this leads, for the 3-dimensional case, to the expansion

$$
\int_\Omega F(\xi) pdf_\xi \mathrm{d}\xi \approx w_0 \cdot F(GF[0]) + w_1 \cdot \sum F(GF[\pm\nu]) + w_2 \cdot \sum F(GF[\pm\nu, \pm\nu]). \tag{2.35}
$$

From this expansion, mean value $\bar{Z}$ and variance σ of the result Z is derived:

$$\bar{Z} \approx w_0 \cdot Z_0 + w_1 \sum_{i=1}^{2n} Z_i, \tag{2.36a}$$

$$\sigma^2 \approx w_0 (Z_0 - \bar{Z}) \cdot (Z_0 - \bar{Z})^{\mathrm{T}} + w_1 \sum_{i=1}^{2n} (Z_i - \bar{Z}) \cdot (Z_i - \bar{Z})^{\mathrm{T}}. \tag{2.36b}$$

Furthermore, determination of the moments of Z and for a normally distributed Z, of the PDF of Z would be possible. For further details see Ref. [144] and references therein. Using these variances, Sobol indices are determined for the global sensitivity analysis [145]. Neglecting higher-order interactions, those can be defined as

$$S_i = \frac{\sigma_i}{\sigma}, \tag{2.37a}$$

$$S_{i,j} = \frac{\sigma_{i,j}}{\sigma}, \tag{2.37b}$$

and

$$S_{\mathrm{T}_i} = S_i + \sum_{j=1}^{N} S_{i,j} \tag{2.37c}$$

wherein σ is the variance of Z and σ_i and $\sigma_{i,j}$ are the partial variances, respectively. In general, σ is a measure of the width of the distribution of the output Z of the non-linear equation F defined in Eq. 2.31, as it is a measure of the sensitivity of output Z to the input variables. The partial variance σ_i, related to the variable X_i, is a measure of sensitivity of Z to X_i. The second order partial variance $\sigma_{i,j}$ is a measure to which extend a change of X_i can vanish the impact of a change of X_j on the output Z. Thus, the later considers the interaction of the uncertain parameters X_i and X_j [145]. For a general definition of the variance and the partial variances it is referred to Chapter 15.1.1, p. 323-326 in Ref. [145]. Commonly, a sensitivity analysis is conducted for model reduction, parameter grouping or parameter estimation purposes [97, 130, 98]. Here, sensitivity is used to assess whether deviations of product properties have a significant influence on the product performance.

In Chapter 4, the non-linear function $F(\cdot)$ is a battery model as introduced in Section 2.3.2. The uncertain variables X_i are the normally distributed cell properties and the output variable Z is the cell's discharge voltage at a specified discharge rate and time, respectively state of charge.

Chapter 3

Parameterization and Practical Identifiability [1]

In this chapter, a parameter estimation is conducted for a P2D model of an NMC vs. graphite cell, which is applied throughout this thesis. As gaining trustworthy results from physics-based models requires thorough parameterization, parameter variations and sensitivity analysis are carried out to ensure uniqueness of the estimated parameter set. By this, the practical identifiability is addressed.

To further assess the P2D model's identifiability, a multi-start approach is introduced as structural identifiability tests reviewed in Section 2.4.2 are not applicable for the P2D model due to its complexity and nonlinearity. Eventually, a three-step parameter estimation is conducted using the open cell voltage curves, C-rate tests, and EIS data, measured in a three-electrode setup.

3.1 Mathematical Method

3.1.1 Electrochemical Models

In this chapter, the P2D model is applied to simulate C-rate tests and the single particle model is applied to simulate impedance spectra. Both models are introduced in Sections 2.3.2 and 2.3.3, respectively. Parameters of the models are listed in a Table 3.1. This parameter set is the result of an non-systematic parameter estimation and is starting point of the further in-depth parameter estimation.

3.1.2 Multi-Step Parameter Estimation

Practical identifiability is influenced by the number of uncertain parameters as well as the amount of experimental data. For an unidentifiable model, reduction of the parameter set or conducting further experiments can enable parameter identification. A further approach is multi-step parameter estimation. Therein, not all parameters are estimated simultaneously. The experimental data are split into parts which are only influenced by a subset of parameters.

In this chapter, a three-step parameter estimation routine is applied for parameter estimation. The experiment contains an OCV measurement for step 1, C-rate tests for step 2, and EIS data for step 3. All experiments are conducted in a three-electrode-setup. For details regarding the experiments, it is referred to Chapter 2.1.4, wherein the electrochemical reference system of this work was introduced. PE step 1 identifies static parameters, like the initial concentration of both electrodes, $c_{a,0}$ and $c_{c,0}$, and the

---

[1]This chapter was first published by Laue and Krewer [95].

Table 3.1: Cell parameters used in the applied model.

parameter	symbol	unit	anode	separator	cathode
layer thickness[m]	δ_{el}	µm	55.25	20.0	60.0
porosity[m]	ε_e	–	0.35	0.50	0.40
diffusion coefficient[a]	D_e	$m^2\,s^{-1}$	$7.15 \cdot 10^{-9}$	$7.15 \cdot 10^{-9}$	$7.15 \cdot 10^{-9}$
diffusion coefficient[a]	D_s	$m^2\,s^{-1}$	$9.35 \cdot 10^{-15}$	–	$1.10 \cdot 10^{-12}$
particle size[m]	R_p	µm	11.5	–	5.5
specific capacity[m]	Δc_{max}	$mol\,L^{-1}$	24.9	–	25.4
electronic conductivity[a]	σ_s	$mS\,m^{-1}$	20.4	–	9.9
exchange current density[a]	i_0	$A\,m^{-2}$	1.47	–	198
transference number[l]	t_p	–	0.24	0.24	0.24
charge transfer coefficient[l]	α	–	0.5	–	0.5
double layer capacitance[l]	C_{DL}	$F\,m^{-2}$	0.2	–	0.2

[m]measured
[a]adjusted
[l]taken from Ref. [146]

specific capacity of both active materials, $\Delta c_{a,max}$ and $\Delta c_{c,max}$. The different steps of the applied parameter estimation routine are summarized in Table 3.2. PE step 2, denoted as quasi-static, is related to kinetic parameters, which are only sensitive at non-zero cell currents, like diffusion coefficients of both active materials respectively, $D_{s,a}$ and $D_{s,c}$, exchange current densities, $i_{0,a}$ and $i_{0,c}$, electric conductivities, $\sigma_{s,q}$ and $\sigma_{s,c}$, and electrode tortuosities, τ_a and τ_c, effecting e.g. the ionic conductivity of the electrolyte. This parameters effect the cell performance at a time scale of minutes to hours.

Table 3.2: Steps and parameters of the multi-step PE approach.

PE step	used experiments	adjusted parameters
1, static	OCV	$c_{c,0}$, $c_{a,0}$, $\Delta c_{c,max}$, $\Delta c_{a,max}$
2, quasi-static	C-rates	$D_{s,c}$, $D_{s,a}$, $i_{0,c}$, $i_{0,a}$, τ_c, τ_a, $\sigma_{s,c}$, $\sigma_{s,a}$
3, dynamic	EIS	$C_{DL,c}$, $C_{DL,a}$, $i_{0,c}$, $i_{0,a}$, $\sigma_{s,c}$, $\sigma_{s,a}$

The electrochemical cell responds at small time scale of seconds and below, is considered in PE step 3. Here, the dynamic cell behavior is analyzed applying impedance spectroscopy. Therein, double layer capacitances, $C_{DL,a}$ and $C_{DL,c}$, are estimated. Exchange current densities, $i_{0,a}$ and $i_{0,c}$, and solid phase conductivities, $\sigma_{s,c}$ and $\sigma_{s,a}$, are recalculated, wherein the result of PE step 2 is used as starting value.

As kinetic parameters do not affect the simulated open full cell voltage curve E_{OCV} vs. discharged charge Q, static parameters θ_1^* are identified using solely the OCV measurement

$$\theta_1^* : F(\theta_1^*) = \min_{\theta_1 \in \mathbb{R}_1^N} F_1(\theta_1) \tag{3.1a}$$

$$F_1(\theta_1) = \sum_{i \in \{\text{an, cath}\}} \left(\sum_{k=1}^{n} \left(\frac{E_{\text{OCV,sim},i}(Q_k, \theta_1) - E_{\text{OCV,exp},i}(C_k)}{\max(E_{\text{OCV,exp},i})} \right)^2 \right. \tag{3.1b}$$
$$\left. + \sum_{k=1}^{n} \left(\frac{Q_{\text{sim},i}(E_{\text{OCV},k}, \theta_1) - Q_{\text{exp},i}(E_{\text{OCV},k})}{\max(Q_{\text{exp},i})} \right)^2 \right)$$

for 2 electrodes i and n sample points k in equidistant potential, respectively charge steps. Deviations in voltage and charge direction are considered to ensure accordance for all SOCs. E.g., at intermediate SOCs the OCV curve is flat and a deviation in charge is large compared to a deviation in capacity. In contrast, at low SOC at a steep OCV curve deviations in voltage are dominant. Summing up both deviations leads to a high accordance for the entire SOC range. Further, this allows to identify the intercalation steps of the graphite anode precisely.

In PE step 2, static parameters are not adjusted anymore. Hence, a smaller subset of parameters θ_2^* is estimated from the C-rate tests with m different C-rates j

$$\theta_2^* : F_2(\theta_2^*) = \min_{\theta_2 \in \mathbb{R}_2^N} F_2(\theta_2) \tag{3.2a}$$

$$F_2(\theta_2) = \sum_{i \in \{\text{an, cath}\}} \sum_{j=1}^{m} \left(\sum_{k=1}^{n} \left(\frac{U_{\text{sim},i,j}(C_k, \theta_2, t) - U_{\text{exp},i,j}(C_k, t)}{\max(U_{\text{exp},i})} \right)^2 \right. \tag{3.2b}$$
$$\left. + \sum_{k=1}^{n} \left(\frac{C_{\text{sim},i,j}(U_k, \theta_2) - C_{\text{exp},i,j}(U_k)}{\max(C_{\text{exp},i})} \right)^2 \right),$$

for 2 electrodes, i, and n equidistant sample points k.

In PE step 3, EIS is simulated at 50 % SOC. The least-square formulation of this step is as follows

$$\theta_3^* : F_3(\theta_3^*) = \min_{\theta_3 \in \mathbb{R}_3^N} F_3(\theta_3^*), \tag{3.3a}$$

$$F_3(\theta_3^*) = \sum_{i \in \{\text{an, cath}\}} \left(\sum_{k=1}^{n} \left(\frac{\text{Re}(Z_{\text{sim},i,k}(\theta_3)) - \text{Re}(Z_{\text{exp},i,k})}{\max(\text{Re}(Z_{\text{exp},i,k}))} \right)^2 \right. \tag{3.3b}$$
$$\left. + \sum_{k=1}^{n} \left(\frac{\text{Im}(Z_{\text{sim},i,k}(\theta_3)) - \text{Im}(Z_{\text{exp},i,k})}{\max(-\text{Im}(Z_{\text{exp},i,k}))} \right)^2 \right)$$

for 2 electrodes, and n frequencies in the experiment.

EIS enables to consider the different time constants of the dynamic processes, and thus allows to distinguish between parameters related to fast electrochemical reactions and slow diffusion processes. Parameter estimation with impedance data leads to further unknown parameters like double layer capacitances of both electrodes, which have a negligible impact in C-rate tests, but a significant impact on the impedance spectra. Hence, PE step 3 provides additional information at the cost of additional sensitive parameters and the need for a model which is able to simulate EIS. Due to that, and due to computational efficiency, a single particle model is used to simulate EIS. This model is described in Section 2.3.3. Where applicable, the same parameters are used as in the P2D model. For the SEI, there are additional parameters which are taken from Ref. [69] as the SEI is beyond the scope of this dissertation.

As the measured charges and voltages have different magnitudes, deviations in the least-square formulations $F_j(\theta_j), \forall j \in \{1,\ 2,\ 3\}$ in Eqs. 3.1 to 3.3 are normalized. Further, the parameter vectors $\theta_j, \forall j \in \{1,\ 2,\ 3\}$ are normalized vs. their starting values due to numerical reasons.[2] For exchange current densities, an exponential transference function is applied between the value chosen by the optimization algorithms and the value used in the model, as the sensitivity of the exchange current density is known to be smaller than the sensitivity of e.g. diffusion coefficients.

3.1.3 Multi-Start Parameter Estimation

Direct approaches to assess the identifiability of model parameters are introduced in Section 2.4.2. Those methods require a closed formulation of the model equation. The P2D model, however, does not satisfy this requirement. A solution to get a closed formulation could be a drastic simplification of the model and linearization. But as it is stated in literature that the nonlinearity bears a significant amount of information which enhances the identifiability of the model parameters significantly [73], an indirect, sample-base, approach is chosen: multi-start parameter estimation[3]. In this approach, different starting points $\theta_{m,0}$ in the domain of model parameters Ω_m are chosen for the parameter estimation algorithm. The domain Ω_m is limited by physical plausible values $x_{lb,j}$ and $x_{ub,j}$ provided by literature.

The choice of the starting points can be guided by different approaches. The parameter space could be sampled in a equidistant mesh. This would lead to a large number of sample points and to many unlikely parameter combinations. Random sampling could reduce the number of sample points, but would lead to non-deterministic results. Further, a small number of parameter vectors could be chosen by physical insight. This could reduce the number of chosen unlikely parameter combinations, such as that all parameters are at the lower boundary, which would lead to negligible dischargeable capacity. To minimize computational cost, this approach is applied in the following.

First, out of nine parameters of PE step 2, the six most sensitive parameters are chosen, namely: solid phase diffusion coefficient of both electrodes, electric phase conductivity of both electrodes, and the exchange current densities of both electrodes. Starting values are varied in the following for those parameters.

For each dimension j of the parameter space Ω_m three values are defined: the boundaries $x_{j,lb}$ and $x_{j,ub}$ and the reference starting value $x_{j,ref}$. From those three values three potential starting points $x_{j,l,0}, \forall l \in \{-,\ 0,\ +\}$ are derived:

$$x_{j,-,0} = 10^{\left(4 \cdot \left(\log(x_{j,\mathrm{ref},0}) + \left(\log(x_{j,\mathrm{lb}})\right)\right)/5\right)}, \tag{3.4a}$$

$$x_{j,0,0} = x_{j,\mathrm{ref}}, \tag{3.4b}$$

---

[2]Common least-square algorithms for nonlinear problems will choose there step size related to the smallest parameter, e.g. a diffusion coefficient. Applying this step to a parameter which is orders of magnitude bigger, e.g. solid phase conductivity, the model seems non-sensitive to this parameter due to the small step.

[3]Multi-start approaches are known for global optimization and can be considered to be state-of-the-art.

$$x_{j,+,0} = 10^{\left(2\cdot\left(\log(x_{j,\mathrm{ref}})\right)+\left(\log(x_{j,\mathrm{ub}})\right)\right)/3}.$$ (3.4c)

The applied logarithm gives weight to the fact that literature values for some parameters range over several orders of magnitude, and thus the starting values should do as well. For clarity, consider the example of $x_{\mathrm{lb}} = 10^{-5}$, $x_{\mathrm{ref},0} = 1$, and $x_{\mathrm{ub}} = 3$. Choosing samples on a linear scale (arithmetic mean) would, for instance, lead to $x_{-,0} \approx 0.5$, $x_{0,0} = 1$, and $x_{+,0} = 2$. While $x_{+,0}$ is located reasonable, $x_{-,0}$ neglects entirely the magnitude of the lower bound. Applying the logarithmic scale introduced above, we get: $x_{-,0} \approx 0.056$, $x_{0,0} = 1$, and $x_{+,0} \approx 1.44$. Thus, the starting value $x_{-,0}$ is a magnitude lower than before, stretching the starting values to a wider sub-domain of the parameter space Ω_{m}. The sample generator function, Eqs. 3.4a to 3.4c, and weightings therein should be adjusted to the individual parameter estimation problem.

As three potential starting values for six parameters would lead to a total of 729 different starting points, promising starting points for PE step 2 are chosen by combination of following rules:

- At least one parameter is set to $x_{j,-,0}$,

- At least one parameter is set to $x_{j,+,0}$,

- At least three parameters are set to $x_{j,0,0}$.

This reduced the number of starting points to 151. The reference starting point $x_{j,0} = x_{j,0,0}, \forall j$ is added as starting point number 152. Further, starting points will be rejected if the discharge capacity at 0.5C is below 50 % of the experimental value. If an starting point is not rejected, parameter estimation step 2 is conducted as introduced in Section 3.1.2.

3.2 Results and Discussion of Identifiability of the P2D Model

In this section, the P2D model is parameterized and its practical identifiability is addressed.

3.2.1 Unidentifiability of the P2D model from C-rate tests

Multi-start parameter estimation was conducted for PE step 2 with sample points as introduced in Section 3.1.3. Regarding identifiability, PE steps 1 and 3 are of minor interested and are discussed in Appendix A. For further discussion of the P2D model's unidentifiability in PE step 2, two parameter sets with an excellent accordance between the experimental discharge curve and the simulation but significant differences in the parameters are hand-picked[4]. The two parameter sets are listed in Table 3.3. The corresponding simulated discharge curves are shown in Figure 3.1. Differences between the curves are negligible and below the typical experimental accuracy. The two different parameter sets vary significantly regarding the cathode properties. The cathode solid phase diffusivity is decreased between the first and second parameter set by a factor of

---

[4]For the sake of completeness, all parameter estimations resulting from the different starting points defined in Section 3.1.3 are discussed in Appendix B.

Table 3.3: Exemplary parameter sets from multi-start parameter estimation.

parameter	$D_{s,a}$	$D_{s,c}$	$\sigma_{s,a}$	$\sigma_{s,c}$	$i_{0,a}$	$i_{0,c}$
unit	$\mathrm{m\,s^{-2}}$	$\mathrm{m\,s^{-2}}$	$\mathrm{S\,m^{-1}}$	$\mathrm{S\,m^{-1}}$	$\mathrm{A\,m^{-2}}$	$\mathrm{A\,m^{-2}}$
solid lines	1.17×10^{-14}	6.72×10^{-13}	81.3	25.0	0.38	1.36
dashed lines	1.23×10^{-14}	3.47×10^{-13}	0.68	0.055	0.37	2.40

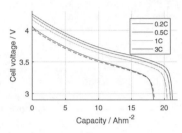

Figure 3.1: Discharge curve of model with parameter set 1 (solid lines) and 2 (dashed lines)

two, solid phase conductivity by a factor of 450, and the exchange current density is increased by a factor of 70. At the anode, only the solid phase conductivity is changed significantly.

Further, a 2D parameter variation is conducted wherein cathode exchange current density is varied (first dimension) and cathode solid phase conductivity and simultaneously cathode solid phase diffusivity (second dimension) are varied simultaneously. It is ensured that both parameter sets in Table 3.3 are a sample of this parameter variation. Results of the parameter variation are shown in Fig. 3.2. It displays $1 - \tilde{F}_2(\theta_2^*)$ for all samples, wherein $\tilde{F}_2(\theta_2^*)$ is the the normalized residuum. The parameter variation reveals the unidentifiabilty of the P2D model with C-rate test data, as an infinite number of solutions is visible. Thus, the combined impact of diffusion and conduction in cathode particles is not distinguishable from the impact of the surface reaction. To overcome this limitation, EIS data are used for the model identification in the following section. The different time constants of reaction and transport processes are used for separation of cathode exchange current density and cathode solid phase conductivity.

To further illustrate the cause of unidentifiability, a sensitivity analysis is carried out for the two different parameter sets, both providing a optimal accordance of simulated and measured C-rate test. The analysis output parameter is the residuum F_2. Sobol indices are determined applying a nested point estimate method which was introduced in Section 2.5.2.[5] Results are displayed in Figure 3.3. The two cases, listed in Table 3.3 and illustrated in Fig. 3.1, are investigated. The model is most sensitive to $D_{s,a}$ and $\sigma_{s,c}$ or $i_{c,0}$ in dependence on the case. This reveals the transition between the two arcs of the L-shaped plot in Fig. 3.2, where one arc relates potential losses to the surface overpotential of the cathode, and the other arc relates it to ohmic losses in the cathode. Sobol indices of $D_{s,c}$ and $i_{a,0}$ are marginal and for $\sigma_{s,a}$ they are zero for both cases. Due to that, the

---

[5]PEM sample points were provided by Xiangzhong Xie. The author gratefully acknowledge this.

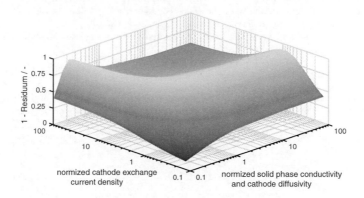

Figure 3.2: 2D parameter variation to illustrated unidentifiability of the P2D model from C-rate tests with three estimated parameters. The z-axis show one minus residuum. Thus, a value close to one is related to high accordance between simulation and experimental data.

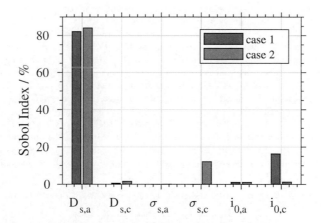

Figure 3.3: Sobol indices for both parameter sets from Table 3.3. Point estimate method was applied for six normally distributed uncertain parameters.

significant changes of anodic solid phase conductivity between the two parameter sets does not effect the discharge curve of the cell, respectively the residuum F_2. Further, this leads to an unidentifiability of the anode conductivity due to non-sensitivity. This further explains the large difference of anode conductivity in Table 3.3.

To enable identifiability of the anode conductivity, a further experiment could be added or the conductivity could be measured ex-situ.

3.2.2 Identifiability through EIS and C-rate Analysis

Electrochemical impedance spectroscopy reveals real and imaginary parts of the battery impedance Z at different frequencies of the sinusoidal current or voltage input signal. At

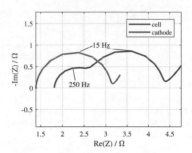

Figure 3.4: Simulated impedance data at room temperature and 50 % SOC.

different frequencies, the impedance is governed by different processes and semicircles are related to electrochemical reactions at particle surfaces or in the SEI [27].

In Fig. 3.4, the simulated impedance spectrum of the lithium ion battery at 50 % SOC is shown, which was investigated with C-rate tests in the previous section. The cathodic impedance contains a single semicircle with a characteristic frequency of 15 Hz and a positive non-zero real part at an imaginary part of zero at high frequencies. This point is related to the conductivity of electrolyte in cathode and separator. The full-cell impedance spectrum contains a second semicircle at high frequencies (250 Hz). This can be related to the SEI [27].

Applying the SPM and PE step 3, solid phase conductivity $\sigma_{s,c}$ and exchange current density $i_{0,c}$ are estimated as listed in Table 3.4. The residuum of the objective function is 0.0038. Simulation and experimental data of the identified cathode are shown in Fig. 3.5. Also listed in Table 3.4 are reference values from literature, which are qualitatively

Table 3.4: Cathode conductivity and cathode exchange current density estimated from impedance data.

Parameter	Estimated value	Literature value	Ref.
$\sigma_{s,c}$ in $S\,m^{-1}$	0.18	0.11	[27]
$i_{0,c}$ in $A\,m^{-2}$	7	1.20	[98]

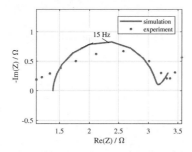

Figure 3.5: Simulated and experimental impedance data of the cathode at room temperature and 50 % SOC.

in good accordance to the value determined based on simulations. It should be pointed out that Heins and Schröder used exactly the same electrodes as in this work [27], while Vazquez-Arenas had different electrodes but at least the same active material [98]. Plotting this parameter combination of $\sigma_{s,c}$ and $i_{0,c}$ in the 2D parameter variation of the

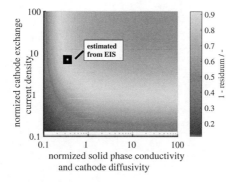

Figure 3.6: Position (black square) of estimated parameters from EIS data in the 2D parameter variation of the C-rate test. Parameter variation data are equivalent to Fig. 3.2.

previous section, we get Fig. 3.6.

The parameters derived from EIS are in good accordance with one of the infinite parameter sets derived from the C-rate test. Results of Bizeray et al. show the structural identifiability of conductivity and exchange current density for the combination of a SPM and EIS [113], even so the entire SPM is not identifiable from EIS data. Hence, the parameters of the cathode, which are required to simulate the discharge behavior in a C-rate test, can be identified uniquely for the applied combination of C-rate test and EIS. Unidentified remain the conductivity of the anode, as it is non-sensitive as shown in Fig. 3.3.

3.3 Concluding Remarks about Identifiability of the P2D

In conclusion, unique parameters of the P2D model were identified successfully through combination of OCV, C-rate, and EIS data in a three electrode setup. The classical C-rate test did not provide sufficient information for model identification and the anodic solid phase conductivity is unidentifiable due to non-sensitivity, even so the combination of C-rate tests and P2D models was the most common case in Table 2.3.

Beside the parameterization of the P2D model, which will be used in the following chapters, the conducted investigations lead to some general findings regarding the parameterization of models and show the limitations of modeling. First, published results of parameter estimation have to be treated with suspicion since often a large number of parameters is estimated, and uniqueness and sensitivity is not addressed. Second, parameterization of an electrochemical model requires generally a combination of dynamic and constant current measurements. Further, without proof of the accuracy of the parameter set, depicted insight into the states is of minor interest as well.

The parameterization conducted in this work can provide guidance for future researchers to conduct a proper parameterization of their models. It gives helpful advice and sensitizes the issue of non-uniqueness.

Chapter 4

Uncertainty Quantification for Large-scale Battery Production

In this thesis, two approaches are used which potentially allow process optimization of battery production. In Chapters 5 and 6, discrete production steps are investigated, while in this chapter the uncertainty propagation of the electrode production towards electrochemical performance is quantified. An uncertainty quantification could allow knowledge-based process designing, e.g. through adjusting individual quality criteria with respect to the quantified sensitivity of the product to the individual quantity. For instance: if a parameter is know to be sensitive and thus critical with respect to cell performance, it has to be monitored attentive to ensure a constant product quality.

Thus, in this chapter an uncertainty quantification is carried out. Experimental deviations are compared to the uncertainty propagation predicted by simulations and the applicability of a nested point estimate method to quantify uncertainty propation applying a P2D model is assessed. This chapter was first published in a journal article related to this disseration. See Ref. [26]. In Section 2.5.1, preliminary work of uncertainty quantification of LIB properties is reviewed. Further, four kinds of property deviation are introduced: particle level deviation, small scale process deviations on single sheet level, termed sub-cell deviation, cell-to-cell and lot-to-lot. The review points out that the combined effect of the first three kinds of deviations have only been the subject of simulation studies implicitly, in so far as real-3D structures were considered, but not as an effect of the production process. Those models are computationally time-consuming and prone to numerical issues. In addition, effects of single parameter distributions are indistinct and an unambiguous assignment is not possible.

In this chapter, a mathematical model is introduced that is feasible to simulate process deviations on a sub-cell level (2nd kind). Parameter distributions of the 1st kind could easily be included in the model, but are beyond the scope of this chapter which focuses on a novel approach to model distributions of electrode properties on a sub-cell level. The introduced model's computational cost is significantly lower than those of real-3D models. For classification of models and their computational cost it is referred to the review of Ramadesigan et al. [147] and Section 2.3.1 in this work.

Applying this model to an exemplary lithium-ion battery, global sensitivity analysis and uncertainty quantification are presented. This analysis allows and in-depth understanding of production induced effects on the cell internal processes and cell performance. This enables a future knowledge-based optimization of the battery production.

4.1 Mathematical Model

In this section, the methodology applied in this work is introduced. A given lithium-ion battery, containing state-of-the-art active material, is simulated by a physics-based model and the model is evaluated using a sample-based UQ approach.

The applied electrochemical model as well as the point estimate method are introduced in Sections 2.3.2 and 2.5.2. The electrochemical reference system is described in Section 2.1.4 and the applied model parameters are listed in Table 3.1. In the following, additional methods and model extension are introduced.

4.1.1 Uncertainty Quantification using Monte Carlo Simulations

Monte Carlo simulations are commonly used for uncertainty quantification. MC simulations generate samples from the probability distribution via random inputs using different sample techniques and estimate the statistical information of the model output. Consider a model $F(X_1, X_2, \cdots, X_N) \in \mathbb{R}^N$. From the parameter space for $(X_1, X_2, \cdots, X_N)$ a finite number m of sample points $\xi_i = (X_{1,i}, X_{2,i}, \cdots, X_{N,i})$, $\forall i \in \{1, \cdots, m\}$ are drawn randomly. The model is evaluated leading to m simulated observations of the model output: $Z_i = F(\xi_i)$, $\forall i \in \{1, \cdots, m\}$ which can be evaluated in terms of mean value and variance if a sufficient number of samples is drawn. The method is straightforward for implementation and can provide accurate estimation. In the context of Batteries, MC was applied e.g. by Mendoza et al. [148] and López et al. [115]. However, the computational demand is often, as in this case, unaffordable since it requires a large numbers of model evaluations to approximate the real statistical information of the model output. Alternatively, the point estimate method can be used as it requires much less model evaluations compared to MC simulations.

In this work, MC simulations are used for cases, where PEM could lead do misleading results, and for comparison and validation of PEM. Therefor, random sample points of deviating electrode properties are generated and C-rate tests are simulated. Due to the large number of sample points this enables the discussion of mean value and shape of probability density function of the model output. This approach was applied e.g. in Ref. [74].

4.1.2 Battery Model

In this section, an extended model based on the P2D model is introduced. It also considers distributed electrode properties on a sub-cell level.

In contrast to the homogenized standard model (see Section 2.3.2), deviations of electrode properties along the cell area A_{cell}, i.e. in y- and z-direction, shall be addressed in the extended model, while a homogeneous structure is assumed along x similar to the standard model. Beside the investigated processes which cause mainly deviations in y- and z-direction, there are processes, like drying, which causes non-homogeneities in the x-direction [33, 32]. However, considering deviations in three dimensions would require a full 3D model. This would not be beneficial for the conducted sample-based PEM, due to its high computational cost. Hence, in this work the focus is set on deviations of electrode properties along the cell area.

Therefore, A_{cell} is divided into a finite number of sub-cells A_i of slightly different properties, e.g. solid volume fraction $\varepsilon_{\text{s,i}}$ and electrode thickness $\delta_{\text{el},i}$. Based on the assumptions in the standard model in Eq. 2.6, and neglecting the resistance of the current collector, as it is small compared to the resistance of the electrodes, $\sigma_{\mathrm{Cu}} = 58 \times 10^6\,\mathrm{S\,m^{-1}} \gg \sigma_{\mathrm{C_6}} = 0.11\,\mathrm{S\,m^{-1}} \gg \sigma_{\mathrm{NMC}} = 68 \times 10^{-3}\,\mathrm{S\,m^{-1}}$, see Refs. [149], [109] and [7], respectively, the liquid phase diffusion in the sub-cells can be calculated separately in one single 1D-scheme per sub-cell. This assumption leads to a parallel ar-

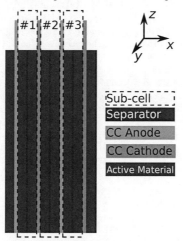

Figure 4.1: Model domains of the extended model.

rangement of sub-cells with independent diffusion processes, but with a uniform voltage and coupled currents, respectively.

There are a few models of battery systems introduced in literature considering cell-to-cell deviations. For instance, Dubarry et al. simulated a battery pack consisting of equivalent circuit models [150, 151]. Kenney et al. applied a serial set of single particle models to simulate cells stacked in sequence [74]. However, the scope of those models was aging prediction or control purposes in battery management systems, and it is far from the scope of this work, which is the quantification of deviations of electrode properties along the cell area and the assessment of the feasibility of the applied PEM for this purpose. In addition, the P2D model applied in this work is superior to an ECM or a SPM in considering physical product parameters and validity for a wide C-rate range.

In addition to the states of the N parallelly coupled P2D models, the current of each area, I_i is an additional state variable. The sub-cell currents have to obey two conditions: conservation of charge and equality of voltage in all sub-cells:

$$0 = I_{\mathrm{total}} - \sum_{k=1}^{N} I_k \qquad (4.1)$$

$$0 = U_{\mathrm{cell}} - U_k\,, \quad k = 1, \cdots, N \qquad (4.2)$$

Using the model introduced above, it is possible to discretize a cell in sub-cells with different electrode properties which can be varied independently for both electrodes.

Both models are implemented in MATLAB 2016a and run on a CentOS Linux desktop PC @3.40GHz and 32 GB of RAM.

4.1.3 Parameters of the Reference Cell

The method introduced above is applied to a LIB containing graphite and NCM111 as anode and cathode, respectively. The properties of the electrode and the cell setup are summarized in Section 2.1.4 and the parameters and their estimation are described in Chapter 3 in great detail. In the further, the parameters considered to be uncertain are addressed.

In this chapter, a parameter is uncertain, if it is effected by deviations of the investigated process steps: coating and calendering. This includes layer thickness and porosity but excludes all intrinsic parameters, like the solid phase diffusion coefficient. In contrast, the solid phase conductivity is a macroscopic effective parameter, which is effected e.g. by the carbon black-binder matrix, which is effected by coating and calendering. Eventually, this classification leads to similar uncertain parameters as common in literature, see Section 2.5.1 in this work. If not noted elsewhere, all uncertain parameters X_i are assumed to be independent and normally distributed with $X_i = \bar{X}_i \pm 5\%$. For deviations observed in literature, see Section 2.5.1 in this work.

4.2 Results and Discussion

The outline of this section is as follows. Deviations of experimental cells and product parameters are summarized. Next, a local and global sensitivity analysis (SA) are carried out. The global SA applying the nested PEM is assessed in comparison to the first order SA. Where appropriate, the influence of sub-cell level deviations is assessed.

4.2.1 Deviation of Product Parameters

Hoffmann et al. investigated process deviations in a batch processes applying 9 A h-pouch cells of identical electrode material as the electrochemical reference system applied in this work [25]. The author of this dissertation published an article containing i.a. an analysis based on raw data of Hoffmann et al. Results of this analysis (see Ref. [26]) is repeated in this subsection in abridged form.

The layer thickness of the evaluated cells has a normal distribution with standard deviations of 1.2 % and 0.88 % for anode and cathode, respectively [25]. With increasing C-rate, the standard deviation is 3.4 %, 4.3 % and 4.6 %. The experimental ohmic cell resistance is 12.5 mΩ ± 4.0 mΩ. The standard deviation of this parameter is 32 %, which is high compared to the other investigated parameters [25].

The deviations of layer thickness and discharge capacity of the evaluated cells in Refs. [25, 26] are in the range of deviations in the literature discussed in Section 2.5.1. Besides, it should be noted that the standard deviation of the discharge capacity is significantly higher than the standard deviation of the layer thickness of anode and cathode, respectively.

In Monte Carlo simulations of Laue et al., deviations as measured and listed at the beginning of this section are considered. This leads to a simulated standard deviation of the discharge capacity at 2.9 V of 0.45% and 0.5% at 0.1C and 2C, respectively. While the increase of the deviation with increasing C-rate is significant in the simulations

(+11.1%) and in the experiments (+35.3%), the width of the simulated distribution with a deviation of ±0.45% to ±0.5% is significantly lower than in the experiments with a deviation of ±3.4% to ±4.6%.

The differences between large cells and laboratory cells as well as the difference between predicted and measured deviations suggests that not all relevant uncertain parameters in the cell production have been identified. Also, deviations in the cell assembly and the electrolyte injection are not quantified yet as imaging the influences of e.g. electrode stacking would require a full-order electrochemical 3D-model. Furthermore it is likely, that there are process deviations that affect parameters, which are not covered by the classical Doyle-Newman model in general. Furthermore, it is possible that the sensitivity of the mathematical battery model to its parameters does not fit the sensitivity of the real battery to its parameters, even so the model is feasible to describe the discharge performance of the reference cell accurately.

Based on these findings, in the further simulations, additional uncertain parameters are considered which are assumed to be also effected by process uncertainties in the production.

4.2.2 Parameter Sensitivity Analysis

In the following, a global first-order sensitivity analysis is carried out to investigate the influence of the different product parameters on the cell capacity. This should allow a quantification of the effect that an input variable like the layer thickness has on an output variable like the discharge capacity. However, this approach is not feasible to quantify any interactions between parameters. To evaluate the influence of all uncertain input parameters, Monte Carlo simulations are conducted. Using the point estimate method, a global sensitivity analysis is conducted to reveal then interactions between uncertain parameters. This method could be feasible to substitute the combination of first-order SA and Monte Carlo simulations for uncertainty propagation. If so, using the PEM could reduce the computational cost for uncertainty quantification significantly. For simplification, an input parameter is denoted as sensitive if there is an output parameter that is sensitive to this input parameter.

First-order Sensitivity Analysis A global first-order sensitivity analysis is carried out by varying each parameter consecutively while all other parameters are kept constant. The varied parameters are layer thickness, δ_{el}, porosity, ε_e, particle size, R_p, and specific capacity, Δc_{max}, of anode and cathode, respectively. Figures 4.2a - 4.2c show the resulting discharge capacities at 0.2C, 1C, and 3C, respectively. Parameter variations are ±10% to investigate a range slightly bigger than the expected range discussed in Section 4.1.3. In general, the slope of these graphs is a measure of sensitivity. In these figures, the graphs of variations of the effective electric conductivity are not shown, as this parameter is in-sensitive for all C-rates for the investigated cell.

Sensitive parameters at 0.2C are the porosities, specific capacity and layer thickness, see Fig. 4.2a. A change of those parameters causes a linear decrease of the discharge capacity, but no significant increase compared to the cell with reference parameters. This is attributed to the good balancing of the electrodes and to the fact that the cell performance is solely limited by the electrode with the lower theoretic capacity $Q_{theo,i}$

$$Q_{theo} = \min(\{Q_{theo,a}, \ Q_{theo,c}\}) \tag{4.3a}$$

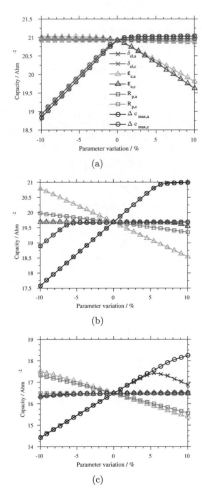

Figure 4.2: First-order SA: Discharge capacity vs. variation of parameter. a) 0.2C, b) 1C, c) 3C. The legend in Fig. 4.2a is valid for all three plots.

$$Q_{\text{theo}} = A_{\text{cell}} \cdot \min(\{\varepsilon_{\text{s,a}} \cdot \delta_{\text{el,a}} \cdot \Delta c_{\text{max,a}}, \ \varepsilon_{\text{s,c}} \cdot \delta_{\text{el,c}} \cdot \Delta c_{\text{max,c}}\}) \qquad (4.3\text{b})$$

as kinetic effects have little influence at 0.2C. As such, also the particle sizes $R_{\text{p},i}$ are insensitive. The slopes of the graphs of both electrodes are similar. The occurrence of the transition between sensitivity and in-sensitivity of many parameters at almost exactly zero parameter variation illustrates the precise balancing of the cell for the operation at low cell currents.

As seen in Fig. 4.2b, at 1C the performance becomes sensitive to the particle size of the anode while the cathode's layer thickness and specific capacity has an effect only when decreasing more than $-5\,\%$. In this range, there is a linear effect on the discharge capacity again. In contrast, the anode's layer thickness and specific capacity are sensitive till a threshold of about $+5\,\%$ after which they are in-sensitive. The sensitivity of the cathode's porosity is negligible at this C-rate. As moderate changes of cathode parameters cannot alter the cell performance, the discharge capacity at 1C is not limited by the capacity of the cathode. The anode layer thickness, $\delta_{\text{el,a}}$, and, specific concentration $\Delta c_{\text{max,a}}$ of the anode shows a similar sensitivity as $\delta_{\text{el,c}}$ and $\Delta c_{\text{max,c}}$ in their sensitive range, respectively. For the investigated range of porosity and particle size of the anode, the system behaves linearly. The sensitivity of the performance to the anode's particle size indicates a limitation of the cell by diffusion processes in the active material of the anode. In summary, kinetic effects become significant as indicated by the increasing sensitivity to the particle size.

The first-order sensitivities at 3C are similar to those at 1C. The sensitivity to the particle size, $R_{\text{p,a}}$, increases while the cathode, in general, becomes less sensitive. A noteworthy feature is a distinguished maximum at an anode layer thickness of about $105\,\%$. This indicates a limitation due to charge or mass transport in the electrolyte phase along the cell. The decrease of the cathode's sensitivity and the distinct maximum of the graph of $\delta_{\text{el,a}}$ suggests higher sensitivities of the anode to high C-rates compared to the cathode. For both electrodes, some graphs show non-linear behavior and transitions between linear areas with different slopes become smoother compared to 0.2C in Fig. 4.2a. This indicates for 3C an interaction of different limitations, that decrease cell performance simultaneously.

From Figs. 4.2a - 4.2c, four types of parameters are derived: first, in- or minor-sensitive parameters, second, parameters sensitive in the entire parameter range (linear and non-linear), third, parameters sensitive till a certain threshold and fourth, parameters sensitive from a certain threshold. The classification of parameters could be used for an advanced optimization or balancing process that considers limitations at different C-rates.

Regarding the design of batteries, the conducted global first-order sensitivity analysis reveals a need for a model-based balancing approach to setup a battery cell that is well balanced at all applied C-rates as the limitation changes with the C-rate. It will also assist to prevent Lithium plating. Regarding the uncertainty quantification of the product parameters of a lithium-ion battery the non-linearity of the system indicates that the discharge capacity of the battery is to a certain extent not normally distributed. For a significant asymmetry of the capacity, the point estimate method could fail to reconstruct the probability density function. Due to that, Monte Carlo simulations are used to investigate the symmetry of the system output in the following section.

Uncertainty Quantification applying Monte Carlo Simulations   In order to evaluate the shape of the PDF regarding asymmetry, in Fig. 4.3 the PDF of the discharge capacities at 0.2C, 1C, 3C, and 4C are shown. They are determined from Monte Carlo simulations with 10.000 sample points considering the 11 uncertain parameters, with $X_i = \bar{X}_i \pm 5\%$, independent of the deviations determined experimentally. A C-rate of 4C is simulated additionally, to confirm the hypotheses of an increase of standard deviation with increasing C-rate. The corresponding standard deviations are 3.20 %, 3.20 %,

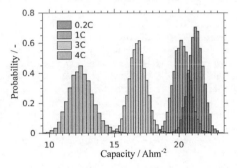

Figure 4.3: Histogram of the discharge capacity of 10.000 simulated discharge capacities at 1C, 3C and 4C considering 11 uncertain parameters with $X_i = \bar{X}_i \pm 5\%$. Sample points are chosen randomly (Monte Carlo Approach).

3.88 % and 7.46 % for 0.2C, 1C, 3C and 4C, respectively. The increase of the standard deviation with increasing C-rate could be related to a change of the sensitivity of the product parameters in dependence on the C-rate and could indicate an interaction of different partially limiting processes and parameters. This leads to the conclusion, that there is a need for a physics-based model to reconstruct process-to-product dependencies, and thus to optimize the battery production. In general, it seems hardly possible to conclude any non-linear dependencies from a normally distributed output.

Global Sensitivity Analysis applying the Point Estimate Method   As illustrated in the first-order sensitivity analysis, the system is significantly non-linear due to the transition of limitations between different processes in the electrodes. Hence, the PEM could, under the condition of high non-linearity or non-differentiability, not be a suitable method. In fact, for the parameters varied in Fig. 4.3 it provides mathematically incorrect Sobol Indexes at 0.2C. As these operational conditions showed the most non-differential behavior in Fig. 4.2a, it is assumed that the failure of the PEM is related to the non-differentiability of the model in this point. However, it shall be used to reveal sensitivities and interactions independently of the given set of electrodes. To enable this and to overcome the limitations of the PEM discussed in Section 2.5.2, the PEM is applied on two different cell setups: One with a 10 % increased layer thickness of the anode, $\delta_{el,a} = 60.78\,\mu m$, and one with a 10 % increased layer thickness of the cathode, $\delta_{el,c} = 66.0\,\mu m$. Due to this change of the layer thickness, the cells are not in the point of the non-linearity at 0.2C, as visible in Fig. 4.2a. Comparison of the sensitivities of both cells can also enable more general conclusions which go beyond the influence of the

simulated cell geometry.

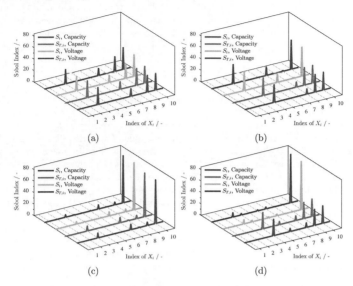

Figure 4.4: Global sensitivity analysis for a cell with a) $\delta_{\mathrm{el,a}} = 60.78\,\mu\mathrm{m}$ at 0.2C, b) $\delta_{\mathrm{el,c}} = 66.00\,\mu\mathrm{m}$ at 0.2C, c) $\delta_{\mathrm{el,a}} = 60.78\,\mu\mathrm{m}$ at 3C, and d) $\delta_{\mathrm{el,c}} = 66.00\,\mu\mathrm{m}$ at 3C. Ten uncertain parameters X_i are considered: 1. $\sigma_{\mathrm{s,c}}$, 2. $\sigma_{\mathrm{s,a}}$, 3. $\Delta c_{\mathrm{max,c}}$, 4. $\Delta c_{\mathrm{max,a}}$, 5. $R_{\mathrm{p,c}}$, 6. $R_{\mathrm{p,a}}$, 7. $\varepsilon_{\mathrm{e,c}}$, 8. $\varepsilon_{\mathrm{e,a}}$, 9. $\delta_{\mathrm{el,i}}$, and 10. A_{cell}. The layer $\delta_{\mathrm{el,i}}$ denotes the layer thickness of the non-adjusted electrode.

C-rate tests are simulated for the two cells at 0.2C, 1C, and 3C. The evaluated output parameters are the discharge capacity and the voltage after discharging 50 % of theoretic capacity at the corresponding C-rate.

Results of these simulations are summarized in Fig. 4.4, where Sobol Indices and Total Solbol Indices are shown for the 10 uncertain product parameters for the influence on capacity and cell voltage at 50 % SOC at 0.2C and 3C. The number of uncertain parameters is reduced by one, as one electrode is over-designed and kept constant to overcome the limitations of PEM regarding the differentiability of the system's output. Each peak indicates a sensitive parameter at the given performance criterion.

Obviously, the cell performance is sensitive to the cell area, A_{cell}, and layer thickness of the non-adjusted electrode, $\delta_{\mathrm{el,i}}$, whereas it is insensitive to electric conductivity of the solid phase, σ_{s}, for both electrodes. Interactions between parameters are marginal as there is no significant difference between Sobol Indices and the corresponding Total Sobol Indices.

Comparison of Figs. 4.4b and 4.4d shows that with increasing C-rate the anode particle size, $R_{\mathrm{p,a}}$, becomes sensitive, which is in good accordance to the first-order sensitivity analysis. Also, Figs. 4.4b and 4.4d reveal that the cell capacity is sensitive to the anode, characterized by $\varepsilon_{\mathrm{e,a}}$ and $d_{\mathrm{el,a}}$, as the cathode's layer thickness was increased, while the cell voltage at 50 % SOC is sensitive to the parameters of the cathode. This is related to the higher slope, dU/dC, of the OCP curve of the cathode compared to the anode. Due to that, a change of the concentration in cathode leads to a significant change of the voltage, even so the limitation at the end of discharge is related to the anode.

In Figs. 4.4a and 4.4c, the sensitivity of cell voltage at 50 % SOC and of the discharge capacity is similar. But at 3C, the cell voltage at 50 % SOC becomes sensitive to the anode particle size, whereas the discharge capacity becomes sensitive to the anode's porosity, compared to 0.2C, respectively.

In general, the most sensitive parameters are the porosity $\varepsilon_{e,c}$ and the specific capacity $\Delta c_{max,c}$, of the cathode, and the layer thickness and cell area of both electrode. The results of the first-order sensitivity analysis for the cell with reference parameters already revealed a strong sensitivity of the battery on the anode. The global sensitivity analysis of a cell with slightly altered electrode layers reveals additionally a sensitivity of the cell voltage to the electrode with the steeper OCP curve independently of the specific cell geometry. In the design of batteries prior to the production process, this has to be considered in the balancing as well as in the set of quality requirements. Thus, regarding large-scale process deviations, the cathode should be produced with higher quality requirements. This supports the common practice to have a slightly overdesigned anode, similar to the cell simulated in Figs. 4.4a and 4.4c. At this instance, quality requirements for the cathode should consider porosity, e.g. effected by calendering, and the intrinsic parameters of the active material which is effected by the material supplier. However, it should be noted that hose guidelines are sensitive to the respective cell and application.

4.2.3 Assessment of the nested PEM

In the uncertainty quantification conducted in this work, two methods have been applied. First, a first-order global SA is combined with Monte Carlo simulations. Second, the nested PEM introduced in Ref. [144] is applied to calculate the Sobol Indexes. Those approaches differ in obtained information about the investigated system, in the computational cost and the limitations regarding non-differentiability of the system.

Comparing Fig. 4.4 with Figs. 4.2a to 4.2c, the feasibility of both sensitivity analysis approaches to reveal sensitive parameters is obvious. Also C-rate dependency of the sensitivity of some parameters could be quantified using both methods.

In addition to the information obtained from the first-order SA, the nested PEM provided the Total Sobol Indices quantifying the interactions between different uncertain parameters. For the investigated system, deviations and sensitivities seem to be insufficiently high to obtain noticeable interactions. A general lack of interactions is unlikely for the investigated system as e.g. Eq. 4.3b provides a lumped approximation of the dependency between cell parameters and cell capacity at a low C-rate. Due to that, it can not be excluded that there could be a limitation of the PEM regarding resolution of minor interactions.

PEM assumes a normal distribution of the output. In contrast, the MC simulations provide additional information regarding the shape of the output distribution. However, as the output is quite symmetrical for the validated operational range of C-rates of 0.2C to 3C, this advantage is of no practical relevance in the here investigated case. It though may be become relevant for another cell chemistry or different operation conditions.

While the results of both approaches are similar, there is a significant difference in the required number of sample points. For the conduced analysis, 10 uncertain parameters are considered in Section 4.2.2 and 5 up to 30 are considered in Section 4.2.4. While the PEM requires $2n^2+1$ sample points, e.g. for MC simulation with 11 uncertain parameters 10.000 sample points are chosen in comparison to the required 243 of the PEM. As such, the nested point estimate method provides comparable results with significantly lower computational costs compared to the approach applying Monte Carlo simulations. A drawback of the PEM is its sensitivity to the differentiability of the investigated system. But this drawback could be avoided if slightly adjusted cells are investigated, as in Section 4.2.2, or a C-rate of 1 or higher is considered, see Figs. 4.2b and 4.2c compared to Fig. 4.2a.

4.2.4 Investigation of sub-cell level deviations

In this section, the point estimate method and the extended model, as introduced in Section 4.1.2, is used to quantify the effect of sub-cell deviations. Five sub-cells, parallel to each other, are considered in the extended model to simulate deviations in the cell due to the production process. Per sub-cell six independent uncertain parameters, $X_i = \bar{X}_i \pm 5\%$, are considered: $\delta_{el,j}$, $\varepsilon_{e,j}$ and $\Delta c_{max,j}$, wherein j indicates anode and cathode, respectively. This results in a total of 30 uncertain parameters. The choice of parameters is based on the high sensitivity of the cell performance to those six parameters, revealed in Section 4.2.2. For comparison of the model considering sub-cell deviations, the same model is used with five identical sub-cells containing thus only six uncertain parameters in total. Due to this, the latter model's discharge curves and PDFs are identical to the results of the standard model.

In Fig. 4.5, the discharge curves of the extended model at 0.2C and 3C and the PDFs of both models at 0.2C and 3C are shown. The PDFs are determined using the PEM. The discharge curves represent the reference cell with five identical sub-cells and

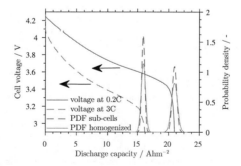

Figure 4.5: Discharge Curves at 0.2C and 3C of the reference cell simulated with the extended model. PDFs of the discharge capacity at 2.9 V of the extended model considering sub-cells and of the model with homogenized membrane properties. Standard deviations are $\pm$ 5 % for all uncertain parameters.

standard parameters. This reference simulation is identical for both models as only the expectation values are used. See the first sample point, ξ_1, in Eq. 2.32.

Comparing both models, there is a marginal shift of the determined mean discharge capacity from $15.93\,\mathrm{A\,m^{-2}}$ and $15.95\,\mathrm{A\,m^{-2}}$ with and without consideration of sub-cell level deviations, respectively, and a significant shrinking of the confidence interval for the model considering sub-cell level deviations, $\sigma = 0.066$ compared to 0.133 for the standard model. Note that the mean discharge capacity is slightly smaller than the discharge capacity of the reference cell, which is $16.16\,\mathrm{A\,m^{-2}}$. The decreased size of the confidence interval seems to be related to the interaction between different sub-cells and balancing of differences in performance as will be shown below.

In order to illustrate the influence of sub-cell variations and the interaction between sub-cells, Fig. 4.6 shows exemplary, the current distribution between sub-cells at one PEM sample point at 3C. In the applied parameter set, the anode layer thickness and the porosity of the corresponding cathode of one sub-cell (sub-cell 4) are increased to $71.51\,\mathrm{\mu m}$ and $36.01\,\%$, while the overall loading is constant in all sub-cells. The other four sub-cells have the parameters of the reference cell. During the first $0.2\,\mathrm{h}$ of operation, all

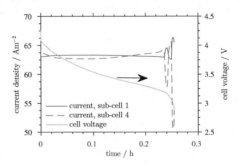

Figure 4.6: Current distribution between sub-cells with different electrode properties. Sub-cell 1 is identical to the sub-cell 2, 3 and 5. In sub-cell 4, the anode layer thickness is increased to $71.51\,\mathrm{\mu m}$ and the cathode porosity to $36.01\,\%$. Apart from that, standard parameters are used.

sub-cells have a similar current density. When the slope of the discharge curve becomes more negative towards the end of operation, there are significant and rapid changes of the current distribution between the different sub-cells on a small time scale of tens of seconds. The current density in sub-cell 4 increases slightly before it drops significantly after about $0.22\,\mathrm{h}$. The other four areas together balance this drop by an increase in their current density. Eventually after about $0.23\,\mathrm{h}$, the current in sub-cell 4 drops by about $20\,\%$ which is more than the remaining areas can balance on a suitable voltage level, causing a large voltage drop, which reaches the lower cutoff voltage.

The observed dynamics in the current distribution is related to the slightly differing surface concentrations of lithium in the active material of different sub-cells. As the surface concentration in the active material in, e.g., an anode of a sub-cell decreases, its potential losses increase and the cell voltage decreases. To ensure equality of voltage between such a sub-cell and the other sub-cells, the current of this sub-cell decreases. In the battery depicted in Fig. 4.6, the effective ionic resistance of sub-cell 4 is reduced

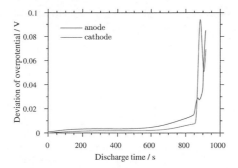

Figure 4.7: Maximum deviation of the overpotential between sub-cells with different electrode properties.

and the active surface and the layer thickness is increased. The latter leads to a higher current density in the first seconds and a lower current in the next about 0.2 h. The increased layer thickness leads to areas in sub-cell 4 which are less utilized within the first 0.2 h. Those areas are utilized when the current density of sub-cell 4 increases above the current density in the other sub-cells.

Parallel to changes of current distribution, there is a high fluctuation of the local overpotentials of up to 100 mV. For illustration, see Fig. 4.7. Therein the maximum difference of the overpotential between the different sub-cells is shown.

The detrimental influence of local high overpotentials regarding cell degradation is discussed in Refs. [60, 152]. In-depth discussion of this is beyond the scope of this work, but it shows that inhomogeneities in cell properties might be a cause for degradation. At low cell currents, a positive effect could be achieved as well: The dynamic balancing could compensate poor performance of some sub-cells and allow a further utilization of the theoretical battery capacity due to the operation at a lower local current density compared to a homogenized cell.

While the results regarding the spatial current distribution can not be validated experimentally in this work, they are highly reasonable. For instance, in experiments of Pastor-Fernández et al. similar current profiles were measured for a parallel set of 18650 cells with different states of health [153]. In addition, similar effects of a non-uniform current distribution were observed in experiments with segmented fuel cells and spatially reduced active area [154, 155, 156]. Also, overpotential overshooting due to non-steady currents was seen in simulations of a direct methanol fuel cell [157]. The findings are also in accordance with simulation results of a battery pack model based on equivalent circuit models [150] and the explanation regarding the detrimental effects of local imbalances or pore blockage to the electrodes leading to lithium plating and accelerated aging due to locally higher current densities [158, 159] and resulting overpotentials.

Summarizing, the studies show that there is an effect of sub-cell level property deviations on the discharge capacity due to non-uniform spatial current distribution. This effect alters the mean discharge capacity and the standard deviation of the cells. Hence, there is a need for additional quality requirements for small-scale variation in the production process, as their influence on local overpotentials is at least as important as the influence of cell-to-cell deviations. For instance, for the investigated stacked pouch cells, it should be avoided to use sheets of different batches and a continuous process should be aimed for.

4.3 Concluding Remarks

A model-based approach for the uncertainty quantification of lithium-ion battery product parameters is introduced which is feasible to evaluate cell-to-cell deviations and sub-cell level deviations by applying an extended Doyle-Newman model. A nested point estimate method is applied for a large number of independent normal distributed parameters, which provides a global sensitivity analysis that reveals a change of sensitivity of the investigated parameters in dependence on the applied C-rate. The method is assessed in comparison with a first-order global sensitivity analysis by parameter variation and a Monte Carlo approach. The PEM is highly efficient regarding computational costs but is limited regarding imaging minor sensitivities. Also, at low C-rates the PEM fails due to the non-differentiability of the system.

The global first-order sensitivity analysis reveals a significant linearity at low C-rates and non-linearity of the investigated electrochemical system at high C-rates. The presence of four types of sensitivities is observed, which have to be taken into account differently in processes like uncertainty quantification or optimization. Application of the introduced UQ approach in large-scale cell production revealed that there is a need to control certain parameters within tighter constraints than others. In general, important parameters are those affecting the theoretical capacity like layer thickness of the limiting electrode. In addition, the discharge voltage is sensitive to the parameters of the electrode that bears the steeper half cell potential curve. In the investigated case, this was the cathode despite the anode causing the limitation of the discharge capacity. For the investigated cell, key parameters for performance are the cathode's specific capacity $\Delta c_{\mathrm{max,c}}$ and the cathode layer thickness, $\delta_{\mathrm{e,c}}$. Hence, the calendaring steps should be monitored in-depth.

The investigation of the influence of sub-cell level deviations revealed that the effect of small-scale deviations is less important for the cell performance of a cell at low C-rates, but it may causes locally higher overpotentials which can be detrimental regarding the long-term performance of the cell, especially at high C-rates. This leads to higher quality requirements for small-scale process deviations within a cell.

The results of this chapter show that a model-based approach is feasible to investigate and optimize the non-ideal LIB manufacturing process to reduce costs and time efforts of the development of e.g. next-generation batteries. In addition, the conducted global sensitivity analysis and the novel model approach reveal electrochemical processes and interactions between product properties and cell performance.

Chapter 5

Micro Structure Modeling of All-Solid-State Electrodes

In Chapter 4, a statistic approach was used to quantify uncertainty propagation between electrode production and cell performance. In Chapters 5 and 6, two discrete production steps, mixing and calendering, are picked out to investigate the optimization potential of those two processes.

In this Chapter, a 3D micro structure model is introduced. It is used to simulate the trade-off between electric and ionic conductivity of all-solid-state electrodes and the influence of the mixing routine. This is of special interest for ASSBs, due to there limited conductivity and complex manufacturing as described in the following section. This chapter was first published in a journal article related to this dissertation. See Ref. [6].

5.1 Preliminary Work

Fundamentals of ASSBs are reviewed in Chapter 2 of this work. Nowadays, the production of ASSBs with porous electrodes is in an early development stage, as badly-connected active material fractions and high overpotentials are major issues [160]. This lead to a focus in research towards material science and thin film electrodes. With the perspective towards e.g. electric cars, however, thick and porous electrodes are getting more relevant. But their total conductivity is still limited and strongly dependent on the production process [12]. Due to that, there are different production processes under investigation: dry processes and those, which require a solvent [161]. Approaches to enhance the processes are e.g. premixing of some components or heat treatment at very high temperatures. For instance, Nam et al. increased the effective ionic conductivity by premixing active material (AM) and solid electrolyte (SE) [161]. The other way around, a premixing of AM and electron conducting additive (ECA) could potentially increase the electronic conductivity.

In general, processing and electrode composition influence the three key parameters of the micro structure: effective electronic conductivity, effective ionic conductivity, and effective active material/electrolyte interface [12]. Hence, composition and processing have to be optimized in parallel to overcome the limitations of today's ASSBs. Modeling is an important tool for such tasks as it promises to yield optimal structures faster than experimental trial and error and as it gives an insight into the limiting processes. Due to that, mixing in ASSB production is chosen as an example to apply the introduced micro structure model.

Relevant modeling approaches in the literature for the investigated objective can be classified into three types: macroscopic homogeneous models like the classical Doyle-

Newman model, 3D micro structure models without electrochemical reactions and, 3D electrochemical models. P2D models [54] allow to simulate e.g. the discharge performance of batteries, with liquid electrolytes as well as solid electrolytes [162], at moderate computational costs. However, due to the homogeneous model structure, the complex influence of the ECA network cannot be addressed in detail. 3D micro structure models can be based on artificial structures [84, 75, 76, 77] or can reconstruct structures from e.g. ccanning electron microscope images [85, 65, 86, 79, 80, 81, 82, 83]. The latter is limited by difficulties to capture the ECA network [81] and has only been applied to LIBs with liquid electrolytes so far. For 3D micro structure models, there are voxel-based models based on the finite element method [65], knot-based discrete element method [84], knot-based resistor network approach models [84], and combinations and coupled versions of those approaches. In general, those models may contain electronic and ionic transport. 3D electrochemical models combine the electrochemical reactions and charge and mass transport processes from the P2D model with the 3D micro structure [85, 86, 81, 65]. Thus, electrochemical processes due to local inhomogeneities can be modeled. Even with domain sizes of very few particle sizes, the price is excessive computational costs [78].

Considering the objective of this work, limits and cost of the different models, a 3D micro structure model with artificial structures is chosen here to investigate the optimization potential of a simulation-based designed ECA network. Optimized structures should lead to decreased ohmic losses and thus improved performance of the ASSB. The most important advantage of the applied model-based approach is the possibility to design new structures, which have not been build yet. The 3D model is compatible to the electrochemical model and would be computationally efficient enough to be included into an optimization routine. As the lack of insight hampers rapid improvement of electrode performance, a 3D structural model would also allow to give insight into optimal structures. Hence, the objective of this chapter is to investigate the influence of the electrode composition and the mixing strategy of the electrode materials on the effective micro structure properties. The investigated model system contains the active material LFP, Carbon Black as electron conducting additive, and a solid polymer electrolyte. Even though most active materials are mixed conductors for electrons and lithium ions, the ionic conductivity of LFP is neglected as it is exceptionally low even compared to other AMs such as LCO or NMC.

In the following, the structure generation algorithm for spherical and irregularly shaped particles and the evaluation of the effective micro structure properties are introduced. Afterwards, parameter studies of the electrode composition are conducted for different processing routines. Case studies are presented. For instance, the influence and the optimization potential of the mixing protocol are shown.

5.2 Computational Methodology

5.2.1 Structure Generation Algorithm

In the following, the applied model is introduced. The main assumptions are summarized in Table 5.2 in the ending of this Section. To flexibly investigate the impact of electrode composition and mixing strategy on the effective electrode properties, voxel-based particle structures are a suitable approach as they allow particles, which are non-uniform in size and shape. To enable non-uniform structures, the particles are based on ran-

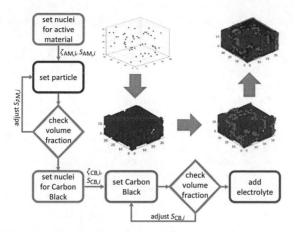

Figure 5.1: Flow chart of structure generation, following the sequence of designing active material structure, then adding the electron conducting additive and then adding electrolyte. Domain and voxel sizes are not representative for the qualitative calculations. The numerical generation algorithm is not related to the production process it self.

domly distributed nuclei. An iterative removal and addition of nuclei allows to control the particle size distribution and the degree of irregularity of the particle shape. The structure generation algorithm is dependent on the model system. For the investigated solid polymer electrolyte, different micro structures can be expected compared to e.g. ceramic solid electrolyte. For the polymer electrolyte the micro structure is assumed to be governed by the active material as this is denser, stiffer and does not melt or partially melt in the production process as the polymer does. This assumption allows a consecutive generation of active material particles first, and later of solid electrolyte. For a ceramic solid electrolyte, active material particles and solid electrolyte particles would be generated simultaneously as they have more similar mechanical properties. In the following, the different steps of the structure generation are summarized. A flow chart of the process is given in Fig. 5.1. The entire simulated three-dimensional space is Ω. The generation process of a structure $\mathcal{M}$ follows a numerical routine to generate a structure which has the following target values: mean particles sizes of AM, $R_{\text{p,AM}}$, and of ECA, $R_{\text{p,ECA}}$; macroscopic volume fraction of AM, ε_{AM}, and of ECA, ε_{ECA}, and the spatial distribution of the ECA. It is not related to the actual production process.

In each structure, four different domains are considered, which resemble the structure of real electrodes: active material, solid electrolyte, an electron conducting additive like carbon black, and the remaining void space. The AM and the SE are assumed to be ideal solids without any intrinsic voids. The ECA is an agglomerate of primary particles which leads to an agglomerate porosity. The pores can be filled with SE or remain void. Both can be found in manufactured electrodes and they depend on the production process. In simulation, the voids can be controlled in the structure generation algorithm.

An iterative approach to determine the AM and ECA particle sizes, $S_{\text{AM},i}$ and $S_{\text{ECA},i}$, is described in the following. In a first step, randomly distributed nuclei are set for the active material domain with a constant nuclei probability in the whole domain Ω, as illustrated in Fig. 5.1. The probability depends on the target volume fraction. Therefor, in preliminary simulations the average volume fraction was determined in dependence on the nucleus probability. This information is used as a lookup table in all further simulations. Subsequently, a spherical particle of random size is built around each nucleus. The particle around each nucleus $\zeta_{\mathrm{AM},i}$ is built up to the respective size $S_{\mathrm{AM},i}$ before the volume fraction of the whole structure is evaluated. Close-by nuclei lead to non-spherical particle shapes and a wider particle size distribution by overlapping of the particles created around neighboring particles. This is a more realistic representation of real electrode structures than spherical particles. To control the volume fraction, the size of the particles $S_{\mathrm{AM},i}$ is adjusted iteratively. All $S_{\mathrm{AM},i}$ are potentially adjusted before the particle generation is repeated for all nuclei. Volume fractions are controlled in a closed loop, while the mean particle size can drift slightly from the target value as the particle size is controlled in an open loop.

The structure is voxel-based. A voxel is a cube of edge length Δx. A structure $\mathcal{M}$ consists of $n_1 \times n_2 \times n_3$ uniform voxels. The numeric particle size S is derived from the geometric particle radius R_{p} and the voxel size by $S = R_{\mathrm{p}}/\Delta x$. Thus, the unit of R_{p} is meter and the unit of S is 1. A nucleus is of the same size as a voxel. Considering a finite number of AM nuclei with locations $\zeta_{\mathrm{AM},i} \in \mathbb{N}^3$ and ECA nuclei with locations $\zeta_{\mathrm{ECA},j} \in \mathbb{N}^3$ with random (equal distributed) particle sizes $S_{\mathrm{AM},i} \in \mathbb{N}$ and $S_{\mathrm{ECA},j} \in \mathbb{N}$, respectively, the structure generation of the electron conducting structure $\mathcal{M}_{\mathrm{el}}$ for any voxel $\varphi = (x_1, x_2, x_3)^{\mathrm{T}}$ with $x_i \in \mathbb{N}$ can be summarized as:

$$
\mathcal{M}_{\mathrm{el}}(\varphi) = \begin{cases} 1 & \text{if } \exists\, \zeta_{\mathrm{AM},i}: \; \|\varphi - \zeta_{\mathrm{AM},i}\| \le S_{\mathrm{AM},i} \; \forall i \\ 2 & \text{if } \exists\, \zeta_{\mathrm{ECA},i}: \; \|\varphi - \zeta_{\mathrm{ECA},i}\| \le S_{\mathrm{ECA},i} \; \forall i \\ & \wedge \nexists\, \zeta_{\mathrm{AM},i}: \; \|\varphi - \zeta_{\mathrm{AM},i}\| \le S_{\mathrm{AM},i} \; \forall i \\ 0 & \text{else} \end{cases} \tag{5.1}
$$

Any vector φ or ζ points to the center of a voxel. In Eq. 5.1, the numbers 1, 2 and 0 are identifier, without a mathematical meaning, for active material, electrode conducting additive and solid electrolyte, respectively. Summarizing Eq. 5.1 for the electron conducting structure, every voxel close enough to an AM nucleus becomes part of the AM particle, every voxel close enough to a ECA voxel becomes part of the ECA particle, if there is no close-by AM voxel, as the active material is considered to govern the micro structure. If there is no close-by AM voxel and no close-by CB voxel, the voxel becomes solid electrolyte/void. In general, Eq. 5.1 forms a spherical around the nucleus voxel. However, due to the interaction of close-by nuclei, non-spherical particles are formed. This overlap is not related to a physical process in an actual structure, but is a method to generate non-spherical particles. For the ion conducting structure $\mathcal{M}_{\mathrm{ion}}$, solid electrolyte and randomly distributed voids $\zeta_{\mathrm{void},i}$ have to be considered:

$$
\mathcal{M}_{\mathrm{ion}}(\varphi) = \begin{cases} 1 & \text{if } \nexists\, \zeta_{\mathrm{ECA},i}: \; \|\varphi - \zeta_{\mathrm{ECA},i}\| \le S_{\mathrm{ECA},i} \; \forall i \\ & \wedge\, \nexists\, \zeta_{\mathrm{AM},i}: \; \|\varphi - \zeta_{\mathrm{AM},i}\| \le S_{\mathrm{AM},i} \; \forall i \\ & \wedge \nexists\, \zeta_{\mathrm{void},i}: \; \zeta_{\mathrm{void},i} = \varphi \; \forall i \\ 0 & \text{else} \end{cases} \tag{5.2}
$$

In Eq. 5.2, the identifier 1 denotes the ion conducting phase, which is assumed to be solely the solid electrolyte. The identifier 0 denotes the ionic non-conducting phase which contains AM, ECA and voids. Here, the ionic conductivity of LFP is neglected. If it should be considered, separation of ECA and AM phases would be required in Eq. 5.2 as it is done in Eq. 5.1. Summarizing Eq. 5.2 for the ion conducting structure, every voxel is part of the ion conducting solid electrolyte, which has no close-by AM voxel, no close-by CB voxel and is no void.

While the generation of the active material domain is straightforward, the subsequent distribution of the electron conducting additive has to consider the production steps of the electrodes, especially the applied mixing protocol. Again, randomly distributed nuclei are distributed in the remaining non-active material space. In contrast to the active material domain, the nuclei probability for the ECA domain is not constant in the electrode. Due to that, $\mathcal{N}$ is introduced, which is equal sized to $\mathcal{M}$ and contains the number of neighboring voxels that are filled with active material for all liquid filled voxels

$$
\mathcal{N}_{i,j,k} = \begin{cases} \sum_{l=i-1}^{i+1,} \begin{cases} 1 & \text{if } \mathcal{M}_{\mathrm{el},l,j,k} = 1 \\ 0 & \text{else} \end{cases} \\ + \sum_{l=j-1}^{j+1,} \begin{cases} 1 & \text{if } \mathcal{M}_{\mathrm{el},i,l,k} = 1 \\ 0 & \text{else} \end{cases} \\ + \sum_{l=k-1}^{k+1,} \begin{cases} 1 & \text{if } \mathcal{M}_{\mathrm{el},i,j,l} = 1 \\ 0 & \text{else} \end{cases} \quad \text{if } \mathcal{M}_{\mathrm{el},i,j,k} = 0 \\ 0 & \text{else} \end{cases} \quad \forall i,j,k \tag{5.3}
$$

wherein $i \in \{1, \cdots, n_1\}$, $j \in \{1, \cdots, n_2\}$ and $k \in \{1, \cdots, n_3\}$. Also, the domain Ω_{ECA} is defined as a sub-domain of Ω, which includes all possible voxels for an ECA nucleus. Depending on the mixing protocol, Ω_{ECA} changes. For instance, an attachment of the ECA to the AM would be likely, if AM and ECA were mixed before the solid electrolyte was added. In this case, represented by Ω_{ECA}^*, the ECA nucleus probability is zero for any voxel that is more than one voxel away from a voxel containing a AM surface site:

$$
\Omega_{\mathrm{ECA}}^* = \{\varphi \in \Omega \mid \|\varphi - \zeta_{\mathrm{AM},i}\| > S_{\mathrm{AM},i} \ \forall i \wedge \mathcal{N}(\varphi) > 0\}. \tag{5.4}
$$

In contrast, mixing SE and AM first, and then adding the ECA would lead to a more homogeneous distribution of the ECA in the SE. In this case, represented by $\Omega_{\mathrm{ECA}}^{**}$, the ECA nucleus probability is constant for the entire non-AM domain:

$$
\Omega_{\mathrm{ECA}}^{**} = \{\varphi \in \Omega \mid \|\varphi - \zeta_{\mathrm{AM},i}\| > S_{\mathrm{AM},i} \ \forall i\}. \tag{5.5}
$$

By applying a spatially varying (Eq. 5.4) or constant (Eq. 5.5) nucleus probability, nuclei of the ECA are set. Afterwards, ECA particles are built around the nuclei. The particle size $S_{\mathrm{ECA},i}$ is adjusted to reach the target volume fraction of ECA.

The remaining void space is randomly filled with electrolyte until the target volume fraction is reached. The remaining voids $\zeta_{\mathrm{void},i}$ are randomly distributed in the entire non-AM and non-ECA domain. The void size S_{void} is one voxel. In dependence on the production process, the pores between the primary particles of the ECA agglomerates are filled with solid electrolyte or remain void. This process is not discretized.

In Section 5.4.3, electrodes with a large particle fraction and a small particle fraction are evaluated. In this case the generator algorithm is called twice to control the volume fractions of both particle fractions independently. The final structure again obeys Eq. 5.1. This approach is used for active material and the conducting additive.

5.2.2 Evaluation of Structures

The voxel-based structures are generated as described in Section 5.2.1 and transformed into a knot-based resistor network. The center of each voxel becomes a knot and each connection to neighboring knots becomes a resistor, with the resistance considering the conductivity of both involved voxels. To enable a connection of the whole surface facing the current collector, respectively separator, a highly conductive current collector is added to the structure. As introduced before, the structure $\mathcal{M}$ consists of $n_1 \times n_2 \times n_3$ uniform voxels. The current collector is a large voxel of size $n_1 \times n_2 \times 1$ which is connected to a complete layer of voxels. This is illustrated in Fig. 5.2. Therein, the positive of the first and last knot are marked. However, the electronic conductivity of the current collector is that high, that there is no significant influence of the position of the knot at the current collector.

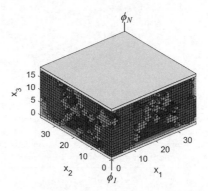

Figure 5.2: Micro structure with attached current collectors and definition of first and last knot. The structure contains AM (large sphericals), ECA (small sphericals), SE (between the particles) and voids (not filled).

A connector conductivity $\mathcal{L}_{i,j}$ in dependence on the voxel size Δx and intrinsic conductivity $\sigma_{i,j}$ is defined as:

$$\mathcal{L}_{i,j} = \sigma_{i,j} \cdot \Delta x. \tag{5.6}$$

The matrix $\mathcal{L}$ is of size $N \times N$, if there are N conducting voxels. The connector conductivity $\mathcal{L}_{i,j}$, between the knots j and i, depends on the present material and on the presence or absence of an particle/particle interfacial resistance, e.g. between the active material and the attached ECA network. Currents between knots flow only in x_1, x_2 and x_3 direction. Thus, diagonal transport is discarded. Ohm's law and Kirchhoff's law allow to calculate the potential in each knot $\underline{\phi} \in \mathbb{R}^N$ by solving the linear equation system:

$$\underline{\underline{\mathcal{L}}} \cdot \underline{\phi} = \underline{I}. \tag{5.7}$$

Therein, the first equation at $i = 1$ is

$$\sum_{j=1}^{N} \mathcal{L}_{1,j}(\phi_j - \phi_1) = I_0 \tag{5.8}$$

wherein I_0 is the applied current at the first current collector. The last equation at $i = N$ is

$$(U_0 - \phi_N) = 0 \tag{5.9}$$

wherein U_0 is the applied constant potential at the second current collector. All intermediate equations are:

$$\sum_{j=1}^{N} \mathcal{L}_{i,j}(\phi_j - \phi_i) = 0 \ \forall i : \ 1 < i < N. \tag{5.10}$$

From the derived potential gradient between two current collectors in a distance of l apart from each other, the effective conductivity $\bar{\sigma}$ of the structure can be derived:

$$\bar{\sigma} = \frac{I_0 \cdot l}{\phi_N - \phi_1}. \tag{5.11}$$

While the math itself is simple, the size of this linear system can become an issue as e.g. a $100 \times 100 \times 100$ voxel structure leads to a size of $\underline{\underline{\mathcal{L}}}$ of $10^6 \times 10^6$. With e.g. about seven non-zero entries per row, direct state of the art solvers in MATLAB consume more than 32 GB of RAM. However, iterative solvers fail due to the big differences between the conductivity of the active material and the conducting additive, leading to non-positive definite matrices. Hence, direct methods are required and the size of the structure is limited by the available RAM to a certain degree. To overcome this limitation, a super-structure approach is used. Smaller voxel-based structures are solved independently, then merged to a super-structure and solved again applying the very same algorithm. For illustration see Fig. 5.3. In the super-structure each sub-structure becomes a single voxel with isotropic homogenized properties. This allows to evaluate structures of relevant size in relatively little computational time and with commonly available RAM. The results of the super-structure are less prone to stochastic effects than e.g. averaging a large number of rather small structures. In contrast to the arithmetic mean value, the super structure provides a physical quantity in dependence on a parallel and serial connection of resistances. Computational time, RAM requirements and stochastic deviations are addressed in the next section.

The applied super-structure approach is well known for different porous systems wherein micro structure properties are addressed and homogenized properties are forwarded to the superimposed continuum model [163].

The evaluation approach derived above is applied for the electronic and ionic conductivity. Beside the effective electronic and ionic conductivity, the effective volume specific

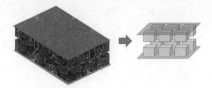

Figure 5.3: Illustration of super-structure approach: discretized structures are evaluated and become a homogenized voxel in the super-structure. Domain and voxel sizes are not representative for the qualitative calculations.

electrolyte/active material interface can be determined by counting solid electrolyte-to-active material interface areas:

$$a_S = \frac{1}{n_1 n_2 n_3 \Delta x} \sum_{i=1}^{n_1} \sum_{j=1}^{n_2} \sum_{k=1}^{n_3} \left(\begin{cases} \mathcal{N}_{i,j,k} & \text{if } \mathcal{M}_{\text{ion},i,j,k} = 1 \\ 0 & \text{else} \end{cases} \right). \tag{5.12}$$

Each interface area is of size Δx^2. In Eq. 5.12, $\mathcal{N}$ is the structure counting neighboring voxels which are active material filled, as introduced in Eq. 5.3, and in $\mathcal{M}_{\text{ion}}$ the number 1 identifies a voxel filled with solid electrolyte. See Section 5.2.1. Any voxel-based reconstruction of a spherical particle has a $4/\pi$ higher surface area than the spherical particle itself.

5.2.3 Physical Model Parameters

Generally, the above described approach is applicable to any ASSB. In this work, an exemplary cell containing LFP as active material, Carbon Black as ECA and PEO plus LiTFSI as solid electrolyte is simulated. The model parameters of this material system is summarized in Table 5.1. The material system is chosen as it is also used in Ref. [31] where the production of ASSBs was depicted in detail. Particle sizes are varied, as they can be adjusted in the electrode production as well and might have a big impact on structures and conductivities. Due to CB agglomerates, the CB particle size can be similar to the LFP particle size [30]. Haselrieder et al. stated a CB agglomerate same of $10\,\mu$m and a CB primary particle size of $40\,$nm. Thus, the here assumed CB particle size can be achieved by controlled deagglomeration. This variants are considered by choosing two sizes of active material and ECA, where one combination, LFP 1 (AM 1) and CB 2, yield particles of identical mean size. For the ionic conductivity an ambient temperature

Table 5.1: Physical model parameters for micro structure simulations.

material	total ionic conductivity	electronic conductivity	particle radius
LFP 1 (AM 1)	$0\,\mathrm{S\,m^{-1}}$	$5.5 \times 10^{-5}\,\mathrm{S\,m^{-1}}$	$3 - 4\,\mu\mathrm{m}$
LFP 2 (AM 2)	$0\,\mathrm{S\,m^{-1}}$	$5.5 \times 10^{-5}\,\mathrm{S\,m^{-1}}$	$1 - 2\,\mu\mathrm{m}$
CB 1	$0\,\mathrm{S\,m^{-1}}$	$500\,\mathrm{S\,m^{-1}}$	$0.5 - 1\,\mu\mathrm{m}$
CB 2	$0\,\mathrm{S\,m^{-1}}$	$500\,\mathrm{S\,m^{-1}}$	$1 - 2\,\mu\mathrm{m}$
PEO + LiTFSI	$5 \times 10^{-4}\,\mathrm{S\,m^{-1}}$	$0\,\mathrm{S\,m^{-1}}$	$0\,\mathrm{S\,m^{-1}}$

of 80 °C is assumed. E.g. see Ref. [31]. The ionic conductivity of LFP is neglected as Amin et al. [164] reported a lithium conductivity of about $3 \times 10^{-7}\,\mathrm{S\,m^{-1}}$ which is about three magnitude lower than the total ionic conductivity of the solid electrolyte. The ionic conductivity listed in Table 5.1 is the total ionic conductivity based on the ion and anion conductivity. This is related to non-polarized composite electrodes. For optimization of the discharge performance of the electrode the low transference number of PEO + LiTFSI would have to be considered in addition. For the electronic conductivity no temperature dependency is considered, as it is marginal compared to the temperature-dependency of the ionic conductivity. For LFP an electronic conductivity 100 times higher than the intrinsic electronic conductivity measured in Ref. [165] is applied, as carbon coated LFP is assumed, due to its higher technical relevance [166, 167]. The electronic conductivity of the ECA is assumed to be $500\,\mathrm{S\,m^{-1}}$. All process variations assume a dry mixing process without binder. A similar production process is investigated in Ref. [31] experimentally.

In the result section, different scenarios are discussed. There the active material volume fraction is varied from 0.4 to 0.8, while the volume ratio between electron conducting additive and active material (ECA/AM) is kept constant.

5.2.4 Numerical Model Parameters

Simulation studies consider an electrode fragment with a thickness of 70 µm and an area of 100 µm × 100 µm. The cubic voxels have an edge length of 1/3 µm. This is similar to 3D FEM models from literature [65] and significantly lower than the voxel size of commonly used imaging methods like scanning electron microscope and computational reconstruction approaches [81]. The electrode fragment consists of 125 sub-structures

Table 5.2: Table of assumptions of micro structure model.

Assumptions
• ionic conductivity of LFP is neglected
• bulk of particles and electrolyte is homogeneous
• All phases have isotropic properties
• All contact areas are ideal contacts
• particle-to-particle-contact areas are not accessible for electrolyte
• Electrode structure is governed by AM phase
• no structure changes along layer thickness
• no ordering of non-spherical particles in any direction
• No particle or ECA breakage
• volume fraction of voids is constant for all electrodes

containing $42 \times 60 \times 60$ voxels, resulting in 18.9 Million voxels. For the applied voxel size, this results in a calculation mesh, respectively sub-structure size, about five time larger than the maximum particle diameter. If the space resolution is increased, respectively the voxel size is decreased, calculation mesh size has to be increased to keep a sufficient large mesh size-to-particle size ratio. The model is implemented in MATLAB 2017b and run on a CentOS Linux desktop PC @3.40GHz with 32 GB of RAM.

5.3 Evaluation of Numerical Effects

A key benefit of the model is its numerical efficiency which should enable it to be applied e.g. in optimization routines. In Table 5.3, the computational cost is listed for three

Table 5.3: Mean computational time for different electrode configurations. ECA/AM ratio of 1/6. Super-structure approach applied.

computational time for	50 % AM, 8.33 % CB	65 % AM, 10.83 % CB	75 % AM, 12.5 % CB
structure generation	5 s	5 s	5 s
sub-structure evaluation	10 s	30 s	41 s
super-structure evaluation	3 ms	3 ms	3 ms

different electrode configurations. The listed time is wall-clock time. The results show an increase with AM volume fraction, as the dominant step is the structure evaluation which is strongly governed by the number of non-zero entries in the structures.

The purpose of using a supercell approach is to reduce computational cost, RAM requirements and the number of sample structures. To assess the feasibility of the supercell to outperform the arithmetic mean, two different approaches are compared. First, 125 sub-structures are merged to a supercell and the conductivity of the supercell is evaluated. Second, only the 125 sub-structures are evaluated and the arithmetic mean value of effective conductivities of all sub-structures is calculated. This provides one estimated conductivity for each approach. To assess the quality of each estimation approach, calculations are repeated ten times with newly generated random-based structures. In Table 5.4, the standard deviation of the calculated electronic conductivity of those ten repetitions is listed. Three different active material volume fractions are considered. Results

Table 5.4: Standard deviation of the stacking approach vs. classical averaging, normalized by mean value. ECA/AM ratio of 1/6.

standard deviation of electronic conductivity	50 % AM, 8.33 % CB	65 % AM, 10.83 % CB	75 % AM, 12.5 % CB
super-structure	1.25 %	53.9 %	3.73 %
average of structures	2885 %	161 %	34.3 %

show a significant reduction of the deviation between the ten instances due to the stacking approach in all cases. Especially, at low volume fractions of AM the arithmetic mean of the 125 sub-structures leads to huge deviations between the ten instances as it is prone to overestimate the impact of large conducting additive clusters. The standard deviation of the ten supercells at 65 % AM is 53.9 %. This deviation is rather high, but is plausible as 65 % AM corresponds to the percolation threshold of the network of the conducting additive. The percolation will be discussed in detail Section 5.4.1. At this point, the electronic conductivity is crucially sensitive to slight changes of the composition. Hence, the deviations between different supercells is high.

In conclusion, the deviations of the stacked super structures are lower than those of the classical averaging, which allows a significant reduction of the sample number and structure size using this method.

To choose a proper super-structure discretization, respectively number of sub-structures, the super-structure discretization is varied, while the voxel size in all sub-structures and a total electrode fragment size, represented by the super-structure, of $70\,\mu$m $\times$ $100\,\mu$m $\times$ $100\,\mu$m is kept constant.

In Table 5.5, mean values and standard deviation of the electronic conductivity of ten instances of supercells with different super-structure discretizations are listed. The

Table 5.5: Mean values and standard deviation of the electronic conductivity applying the stacking approach with different discretization. ECA/AM ratio of 1/6.

super structure discretization	electronic conductivity in $S\,m^{-1}$ with		
	50 % AM	65 % AM	75 % AM
$4 \times 4 \times 4$	$2.9 \times 10^{-5} \pm 1.39\%$	$0.3 \times 10^{-2} \pm 142\%$	$2.1 \pm 3.64\%$
$5 \times 5 \times 5$	$3.0 \times 10^{-5} \pm 1.25\%$	$1.0 \times 10^{-2} \pm 53.9\%$	$2.1 \pm 3.73\%$
$6 \times 6 \times 6$	$3.1 \times 10^{-5} \pm 1.32\%$	$1.1 \times 10^{-2} \pm 109\%$	$2.1 \pm 3.14\%$

results show marginal differences between different super-structure discretizations. For $5 \times 5 \times 5$ super-structures, the standard deviation at 65 % AM is the smallest as it provides a trade-off between large enough sub-structures and a sufficient number of stacked structures. Due to that in the following simulations, this discretization is used.

5.4 Results and Discussion

The objective of this work is to highlight the optimization potential of the mixing and premixing of active material, solid electrolyte and conducting additive. Effects on the effective electronic and ionic conductivity are discussed, as well as to the effective electrolyte-to-active material surface to achieve electrodes with better performance. Throughout this section, effective (electric or ionic) conductivity is referred to the averaged conductivity of the entire structure as described in Eq. 5.11.

The outline of this section is as follows: First, the influence of the ECA-to-AM ratio and the AM volume fraction is investigated for an ECA attached to the AM surface. Here, AM 1 and CB 1 are used. This is followed by a change in the distribution of the ECA in the electrode due to the mixing routine and, then by evaluating the influence of the ECA particle size. All simulations consider one supercell containing 125 sub-structures for each electrode composition.

5.4.1 Influence of Conducting Additive Volume Fraction

The volume fraction of active material is varied for different fixed ECA-to-AM ratios. With decreasing volume fraction of solid electrolyte, the AM and the ECA volume fraction are increasing simultaneously. Here, ECA is completely attached to the AM due to an idealized premixing of ECA and AM. This distribution is illustrated in Fig.

5.4a. The influence of the composition on the effective ionic and electronic conductivity and the active surface area is analyzed.

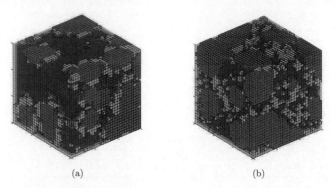

(a)   (b)

Figure 5.4: 3D structures with a) equally distributed ECA ($\Omega_{ECA} = \Omega_{ECA}^{**}$) and b) ECA attached to the AM ($\Omega_{ECA} = \Omega_{ECA}^{*}$). The ECA distribution can be controlled by premixing.

Fig. 5.5 shows the effective ionic and effective electronic conductivity as function of the AM volume fraction. The ionic conductivity decreases with increasing AM volume fraction with a pronounced decrease at 0.7 for an ECA/AM ratio of 1/6 and at 0.75 for 1/9. The electronic conductivity at an ECA/AM ratio of 1/6 increases about 3 to 4 magnitudes at an AM volume fraction of about 0.65 (i.e. 0.11 ECA volume fraction), and the 1/9 ECA/AM curve shows the beginning of a similar drastic conductivity increase at 0.8 which corresponds to 0.09 ECA volume fraction. The electrode without conducting additive shows a slow increase of electronic conductivity and decrease of ionic conductivity. The strong increase of electronic conductivity at an ECA volume fraction of 9 % and 11 %, respectively, is related to the percolation of the ECA network, as from this volume fraction on the highly conducting additives form a percolating network through the entire electrode. Generally, this state should be aimed for, as it provides a low cell resistance and a high utilization rate of the active material. The predicted percolation is a feature of the ECA network, which was also observed experimentally. A general validation of the simulation results of this work is provided in Section 5.4.4 using experimental results from literature. At an ECA/AM ratio of 1/9, the ionic conductivity drops before the percolation threshold of the electronic conductivity is reached at about 80 % AM and 7 % CB. Hence, the remaining PEO volume fraction of 13 % is insufficient to achieve a good ionic conducting network and thus conductivity. In contrast, there is a significant range of active material volume fraction which leads to both, a high ionic and a high electronic conductivity at an ECA/AM ratio of 1/6. This electrode configuration is therefore used as a reference for the further simulation studies.

In Fig. 5.6 effective ionic (solid lines) and electronic conductivities (dots) are shown for a variation of electrode composition. An ECA/AM ratio of 1/6 and zero is simulated. Bruggeman relation are provided for comparison for electronic (dashed gray lines) and ionic conductivity (solid gray line). For electronic conductivity two different Bruggeman-

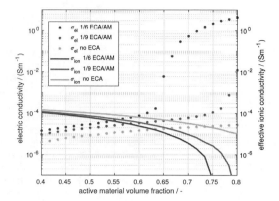

Figure 5.5: Effective electronic and ionic conductivity vs. active material volume fraction for different ECA/AM ratios of 1:6, 1:9 and zero (without ECA). ECA is attached to AM. AM 1 and CB 1 are used.

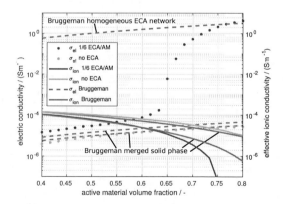

Figure 5.6: Comparison with Bruggeman relation: Effective electronic and ionic conductivity vs. active material volume fraction for different ECA/AM ratios of 1:6 and zero (without ECA). ECA is attached to AM. AM 1 and CB 1 are used.

type relations are considered:

$$\sigma_{\text{el,eff}} = \sigma_{\text{el,AM}} \cdot \varepsilon_{\text{s}}^{2.5} \tag{5.13}$$

and

$$\sigma_{\text{el,eff}} = \sigma_{\text{el,AM}} \cdot c_{\text{AM}}^{2.5} + \sigma_{\text{el,ECA}} \cdot c_{\text{ECA}}^{2.5}. \tag{5.14}$$

In the first, volume fractions of AM and CB are merged to ε_{s}. This classical Bruggeman relation neglects the high electronic ECA conductivity. The later considers a Bruggeman-type term for the AM phase and a second Bruggeman-type term for the ECA network. This approach assumes a homogeneous, respective percolating, ECA network for any non-zero ECA volume fraction. For the ionic conductivity it is:

$$\sigma_{\text{ion,eff}} = \sigma_{\text{ion}} \cdot \varepsilon^{2.5}. \tag{5.15}$$

In the absence of an ECA, the Bruggeman relation is in good accordance with the result of the 3D simulations, for electronic conductivity as well as for ionic conductivity. Also for an ECA-to-AM ratio of 1/6, the ionic conductivity is predicted well by the Bruggeman relation for low AM volume fractions. But at high volume fractions, a crucial decrease of ionic conductivity in the 3D simulations is not reproduced by the Bruggeman relation. This decrease is related to the existence of non-connected pores which are not considered in the homogeneous Bruggeman relation, but can play an important role at low porosities. For the electronic conductivity in the presence of an ECA, the classical relation, Eq. 5.13, is accurate at low AM volume fractions while the relation considering two parallel homogeneous phases, Eq. 5.14, is accurate at volume fractions above the percolation threshold. Those results, validate the 3D model, as well as show the limitations of the Bruggeman relations. For the electronic conductivity the 3D model simulates the transition from Eq. 5.13 to Eq. 5.14, while any of the two relations would not be able to predict the effective electronic conductivity for the entire parameter range.

In Fig. 5.7, the volume-specific active surface area a_{S} is pictured as a function of electrode composition. In general, a_{S} first increases with increasing AM volume fraction, as there is more active material. Here most parts of the AM particles are accessibly for the solid electrolyte. At higher volume fraction, the specific surface area decreases as particle-to-particle contacts between AM particles are becoming more frequent. In addition, the active surface area decreases with increasing ECA/AM ratio, as the the ECA blocks a larger portion of the particle surface. As the ECA is attached to active material, this effect occurs even at low volume fractions. At low active material volume fractions there is a slight increase of surface area with AM fraction for all ECA/AM ratio with a maximum at about 0.6 for the electrode without ECA and at about 0.5 in the presence of ECA. At high AM fractions, there is a decrease of surface area of about 68 %, 47 % and 23 % between the maximum and a volume fraction of 0.8 for ECA/AM ratios of 1/6, 1/9 and zero, respectively. With an increasing ECA/AM ratio, the surface area decreases as the conducting additive blocks the surface and reduces the contact area between solid electrolyte and active material. The optimal volume fraction of active material regarding surface area shifts slightly to lower AM fractions as the ECA/AM ratio increases. Comparing Fig. 5.5 and Fig. 5.7, an active material volume fraction of 0.65 would probably be optimal for an ECA/AM ratio of 1/6 as the electronic conductivity is high, but negative influences of the conducting additive on the ionic conductivity and the SE/AM interface are low.

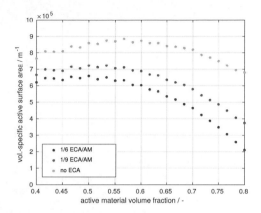

Figure 5.7: Effective electrolyte-to-active material interface vs. active material volume fraction for different ECA/AM ratios. AM 1 and CB 1 are used.

5.4.2 Influence of Spatial Conducting Additive Distribution

Having shown the importance of the electrode composition for the effective conductivity and thus performance, we now discuss the effect of different distributions of the conducting additive, which can be achieved by different mixing routines. Different spatial distributions are investigated, as they would be adjustable in experiments as well. Three scenarios are investigated. First, the ECA is completely attached to the AM due to an idealized premixing of ECA and AM. For illustration see Fig. 5.4a. This scenario is similar to liquid electrolyte LIBs with a solvent-based process. Second, the ECA is distributed evenly as the SE and the AM are premixed, as illustrated in Fig. 5.4b. And third, a three-step mixing routine is applied: A part of the ECA is premixed with the AM, while the remaining ECA is premixed with the solid electrolyte. Eventually both mixtures are combined. This leads to attached conducting additives and to evenly distributed ECA. The third scenario is a combination of the first and second scenario and would be applicable with an additional mixing step. For all simulations of scenario three, half of the ECA is attached to the active material and half is distributed in the electrolyte. A premixing of additive and active material, which attaches the additive to the AM, is used as reference throughout this chapter. All simulations in this section are with an ECA/AM ratio of 1/6.

In Fig. 5.8, the influence of the additive distribution, respectively mixing strategy, on the electronic and ionic conductivity is shown. As can be seen, electronic conductivity decreases with increasing degree of homogeneity. Premixing of ECA and SE shifts the percolation threshold to about five percent points higher volume ratios. The collocation of the additive on the surface area leads to a highly conducting network at lower volume fractions. However, this collocation at the surface reduces slightly the ionic conductivity as it increases the electrode tortuosity, probably by clogging small pores between neighboring active material particles.

The second drawback of the surface collocation is shown in Fig. 5.9. The AM surface

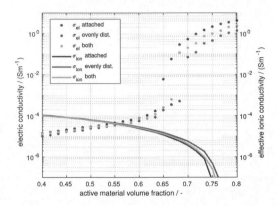

Figure 5.8: Effective electronic and ionic conductivity vs. active material volume fraction for different ECA distributions. AM 1, CB 1 and an ECA/AM ratio of 1/6 are used.

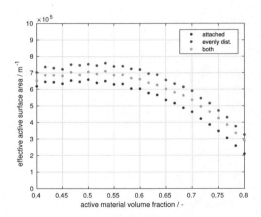

Figure 5.9: Effective electrolyte to active material interface vs. active material volume fraction for different ECA distributions. AM 1, CB 1 and an ECA/AM ratio of 1/6 are used.

area of the electrodes with the attached ECA is about 14 % smaller than the surface area of the electrodes with evenly distributed ECA. Here again, it is shown that the attachment of ECA to the AM surface decreases the surface accessible for solid electrolyte and thus the electrochemical reaction. These simulations show that the mixing strategy has to be optimized for different cells depending on the application, as they probably require different trade-offs between electronic conductivity, ionic conductivity and active surface area.

Summarizing Figs. 5.8 and 5.9, an optimal AM volume fraction would be about 0.65 if the ECA is attached to the AM surface. If the ECA and the SE are premixed, leading to evenly distribution of the ECA, higher volume fractions of AM should be aimed for as ionic conductivity and active surface area decrease less significantly. Here, an AM volume fraction of up to 0.7 could be optimal.

5.4.3 Influence of Size Distribution of Conducting Additives and Active Material

Similar as electrode composition and ECA distribution, the particle size of active material and the agglomerate size of ECA may have a significant influence on the cell performance and could be adjusted in the electrode production. To analyze this effect, we vary the particle size distributions of the conducting additive and the active material, respectively. As listed in Table 5.1, two different conducting additives and active materials are considered, varying in the mean particle size. Applications of the reference materials and of a variation of mixture of both, ECA and AM sizes, are simulated. Mixtures contains 50 vol % of AM 1 and 50 vol % of AM 2, or 50 vol % CB 1 and 50 vol % CB 2. All simulations in this section consider ECA attached to the AM.

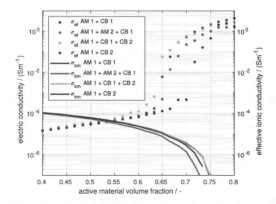

Figure 5.10: Effective electronic and ionic conductivity vs. active material volume fraction for different ECA and AM sizes. ECA attached to AM. ECA/AM ratio of 1/6.

In Fig. 5.10, effective electronic and ionic conductivities are shown for the reference case (blue) and for two additional mixtures, for an ECA/AM ratio of 1/6. All three material combinations have the steep increase of electronic conductivity as shown before. With increasing ECA size, the characteristic step occurs at slightly lower AM volume fractions, while for decreasing the mean AM particle size, it shifts to higher volume fractions. For AM fractions higher than the percolation threshold, the electrode with the reference composition has the highest electronic conductivity. The positive influence of the wide ECA particle size distribution (CB 1 + CB 2) could be related to larger particles building conducting bridges over large pores. At high AM volume fraction, the small ECA could be beneficial as the small particles form a thinner, more efficient network at the active material surface. In contrast to the mixture of CB 1 and CB2, the electrode containing only CB 2 has the lowest electronic conductivity of all simulated electrode composition. This is evidence, that the increase of electronic conductivity is related to the wide ECA particle size distribution and not solely to its mean particle size. As a low percolation threshold with respect to ECA volume fraction has to be aimed for, a mixture of both, CB 1 and CB 2, could be optimal to form a highly conducting network at lower volume fractions. A low percolation threshold of the ECA network would allow to decrease the CB volume fraction and to increase the AM volume fraction to increase capacity without decreasing the solid electrolyte volume fraction and thus ionic conductivity.

Fig. 5.10 also shows that the influence of the CB mixture on the ionic conductivity is negligible. Both curves show the same trends as observed before. The mixture of AM 1 and AM 2 leads to an increased tortuosity, causing an earlier onset of the steep decrease of ionic conductivity.

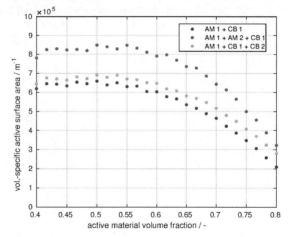

Figure 5.11: Volume-specific active surface area vs. active material volume fraction for different ECA and AM sizes. ECA attached to AM. ECA/AM ratio of 1/6.

In Fig. 5.11, the influence of the choice of additive and active material size on the SE/AM interface is shown. The general trends are as described before. The additive with the smallest particle size reduces the surface area the most, as it can form a thinner coating layer than the bigger particles. However, this influence is quite low with about

10 % between CB 2 and CB 1. The mixture of AM 1 and AM 2 has a significantly increased active surface area, due to the higher surface/volume ratio of small particles. This leads to the conclusion, that an electrode containing AM 1 + CB 1+ CB 2 at an AM volume fraction of 0.65 would be optimal. In addition, this electrode would allow a decrease of the ECA volume fraction, allowing a further increase of the AM volume fraction and thus, volumetric capacity.

5.4.4 Qualitative Comparison with Published Experimental Data

While modeling generally aims to simulate new structures and experimentally not yet evaluated cases, experimental data is required to parameterize or validate models. Therefore, the basic features of the simulation results are compared to published experimental results, where possible. It has to be noticed, that ASSBs with bulk electrodes are in an early development stage yet. Hence, few quantitative data are available. More frequently, research about enhancing the intrinsic properties of active materials, thin film electrodes or polymer and solid electrolytes [168, 169] is published. However, those results are of minor interest for this comparison.

Electrochemical impedance spectroscopy data of Nam et al. suggests an increase of active surface area of about 62 %, of first cycle discharge capacity of about 40 %, and an increase of ionic conductivity by a premixing of the active material and solid electrolyte. Therein, the size of the semicircle in the Nyquist plot was with premixing about half the size than without premixing [161]. Both features are in good accordance with the simulation results in Section 5.4.2 in this work. In the work of Nam et al. the improvement of electrode performance was suggested to be related to an enhancement of the SE-to-AM interface as well to enhancing ionic contacts. This could be related to mechanical or interfacial effects of the SE, as well as on the spatial distribution of the SE, as seen in the results of this work. In addition, a measured AM surface coverage of the SE of 25.1 % and below [161] validates the reduction of the AM/SE interface by ECA and AM/AM contacts, shown in Figs. 5.7, 5.9 and 5.11. Surface blockage due to the conducting additive and binder phase was also observed in EIS data at positive electrodes from Sakuda et al. as with increasing binder amount the first semicircle increased [170].

The percolation threshold of the ECA network, was not reported for ASSBs. However, it is a commonly observed feature in liquid-electrolyte batteries, for instance, seen in calendering studies [171, 172, 173, 84]. Among these, Ott at el. applied DEM simulations and suggest a percolation threshold between 10 % and 15 % volume fraction of carbon black [84].

Further, the suggested optimal electrode composition of about 65 vol % AM, 10 vol % ECA and 25 vol % solid electrolyte is compared to reported electrode compositions in literature. All-solid state NMC cathodes of Sakuda et al. consist of 70 wt % AM and 6 % conducting additive and binder which is close to values used in this work. Also, the NMC was coated with $LiNbO_3$ to increase the surface conductivity [170]. The composition in the work of Nam et al. discussed above, varied between 69 wt % to 84 wt % for the active material and 1.3 wt % to 1.7 wt % for the additives [161]. While there are only few publications about ASSBs revealing the whole composition, it also has to be kept in mind that weight percents are commonly reported, while for modeling, and for the physical phenomena as well, volume fractions are of interest. Hence, a quantitative comparison would require knowledge about the densities of the materials used in literature. However, as the density of e.g. LFP is significantly higher than the density of carbon black, the here suggested optimal composition has shown to be in a range of technical relevance. Also, the reported ASSB compositions contained a binder, which is not considered in the simulations in this work.

5.5 Concluding Remarks

This chapter contributes to understanding and optimization of bulk electrodes for ASSBs via a model-based analysis. The insight is realized by systematically analyzing the impact of the electrode composition and its micro structural properties on the effective electronic and ionic conductivity, and on the electrochemical active area. Various scenarios were simulated with a 3D micro structure model. A super-structure approach allowed to evaluate electrode fragments of $70\,\mu m \times 100\,\mu m \times 100\,\mu m$ with moderate computational costs. The simulated changes in the electrode structure are closely linked to technically relevant scenarios of different mixing strategies. The simulation results highlight the potential to optimize the electrochemical performance and active material utilization rate of all-solid state batteries via structure and composition optimization. Identified design parameters are the particle size and particle size distribution of the electron conducting additive, as well as the premixing of the electrode components. In dependence on the spatial ECA distribution and the ECA particle size distribution, the ECA network percolation occurs at active material volume fraction higher than 60 % for e.g. 10 % ECA volume fraction. In general, premixing of AM and ECA can be used to increase the electronic conductivity. However, every investigated measure to increase the electronic conductivity decreases the ionic conductivity and the effective active surface area. Due to this interaction, a mathematical optimization of the composition and the mixing strategy is of high interest. Such further work requires a coupling of the model introduced in this work with an electrochemical model. See Chapter 6 for the application of this approach for quantification of the calendering effect on the electrochemical cell performance.

Chapter 6

Coupling of Micro Structure and Electrochemical Models

This chapter was first published in Ref. [34]. To quantify the influence of effective micro structure properties on the cell performance an electrochemical model is required. Therefor, the micro structure model, introduced in the previous chapter, is combined with a P2D model, see Section 2.3.2, of a lithium ion battery with liquid electrolyte to investigate the calendering influence. Generally, in this work, the link between electrode structure parameters like porosity and model parameters like effective conductivity is referred to as structure-model parameter relations. The probably most famous structure-model parameter relation is the Bruggeman relation. In the calendering simulations published up to now, effective electrode parameters like conductivity and electrochemically active area, i.e. the material-to-electrolyte interface, have to be estimated for all calendering rates individually [73]. This allows a model based analysis of battery performance based on experimental data, but it causes the model's shortcomings in terms of prediction and optimization.

Comparison of the introduced coupled model of a 3D-micro structure model and an electrochemical P2D model with the classical P2D model and commonly applied micro structure-model parameter relations like the Bruggeman relation allow assessment of the required model complexity for a knowledge-based optimization of the battery production.

6.1 Preliminary Work on Structure-Model Parameter Relations

In electrochemical homogeneous models, structure-model parameter relations provide the link between geometric structure properties like porosity and model parameters like effective ionic conductivity. In the classical Doyle-Newman model three simplified structure-model parameter relations are considered. For the effective ionic conductivity the Bruggeman relation

$$\sigma_{\text{liq}} = \tilde{\sigma}_{\text{liq}} \cdot \varepsilon^{\beta} \tag{6.1}$$

is applied which describes the influence of the porosity ε on the effective conductivity σ_{liq}. The Bruggeman coefficient β is a measure of tortuosity τ, as tortuosity can be determined by:

$$\tau = \varepsilon^{1-\beta}. \tag{6.2}$$

Thus, the bigger β the higher the tortuosity. For an electrode consisting of ideal spheres, β is equal to 1.5. In many publications, it is adjusted to fit simulated cell performance and experiments. In Equations 6.1 and 6.3, and in the further, a tilde denotes a bulk property.

For the electric conductivity, it is assumed:

$$\sigma_s = \tilde{\sigma}_{AM} \cdot \varepsilon_s^{\beta} \tag{6.3}$$

wherein ε_s is the solid phase volume fraction. In some publications the Bruggeman coefficient β is set to one for the solid phase [146] as the Bruggeman relation was developed only for the transport around spherical non-contacting obstacles [174].

For the active surface area, which is the contact area between active material and electrolyte, the estimation

$$a_S = \frac{3 \cdot \varepsilon_S}{R_p} \tag{6.4}$$

is commonly known, which is based on the assumption of mono-disperse particles of size R_p which do not contact each other. Independent on particle-to-particle contacts, the assumption of mono-disperse spheres is limited by the densest possible particle packing of 74 % solid.

In literature, various extended structure-model parameter relations for conductivity exist. Zacharias et al. derived more accurate Bruggeman relations from experimental data for different porosities and Carbon Black amounts to enhance the accuracy of the effective ionic conductivity in an electrochemical battery model [172]. This approach is similar to the one presented in this chapter. Here however, a 3D model is used instead of experimental data, and further structural properties are investigated. Ott et al. introduced a micro mechanical model to derive effective ionic and electric conductivities of structures of mono-disperse spherical [84]. This approach is similar to the micro structure model applied in this work. However, it neglects to consider the crucial impact of Carbon Black. Thus, no model-based approach that considers structural impacts including Carbon Black on the LIB performance is available. The carbon black binder domain was considered by Ngandjong et al. in an electrochemical 3D model [88]. Compared to this model, the coupling approach neglects concentration differences due to spatial effects, but the coupling approach with the P2D model allows optimization due to moderate computation costs. In the work of Bielefeld et al. the commercial software GeoDict was used [175]. The software allows automated generation of particle structures of different particle shapes and the article focused on the contact area of active material and electrolyte and the utilization fraction of active material. Derived percolation thresholds and active surface areas were in good accordance with results derived from the micro structure model in Chapter 5. The major advantage of the self-programmed micro structure model, compared to commercial software, is the flexibility with respect to novel focuses as well as the compatibility with the P2D model, which allows a fluent processing and optimization of batteries. In contrast to the work of Ngandjong et al. and Liu and Mukherjee [88, 87], the applied model is limited to generic distributions, which is not governed by chemical processes in the drying step of the electrode production. Further, Carbon Black and binder are not distinguished. In this contribution we present a study on the crucial role of the spatial Carbon Black distribution on conductivity and active surface area to allow quantification and prediction of the calendering influence.

The outline of this chapter is as follows: From the micro structure model, a set of algebraic structure-model parameter relations is derived which is used to enhance the electrochemical model. Eventually, the feasibility of the framework of enhanced structure-parameter relations and P2D model to reproduce the experimental discharge curves for different calendering rates with a single parameter set is assessed.

This approach is chosen as results of Lenze et al. show that the homogeneous P2D model is able to simulate batteries at different calendering rates if the effective micro structure parameters are known for each calendering rate [73]. Hence, calendering rate dependent modeling of effective micro structure parameters allows an accurate prediction of cell performance at different calendering rates.

6.2 Cell Setup and Test Conditions

For the electrochemical experiments a commercial three-electrode setup was used.[1] PAT-Cells from the EL-Cell GmbH provide a three-electrode setup with a round electrode with a diameter of 18 mm and a separator with an included lithium reference electrode. The separator is a glass fiber separator by EL-Cell GmbH (EU1 000210 0/X). The electrode compositions is summarized in Section 2.1.4. The electrodes were produced in the BatteryLab Factory Braunschweig on a pilot-plant production line and were also studied in Refs. [72, 176, 27]. Ref. [72] provides further information on the respective compression forces of the different calendering rates and Ref. [176] provides details on the production process regarding e.g. the mixing process. The porosity used in this chapter is calculated, based on measured weight (scales XS205, Mettler Toledo) and thickness (micrometer screw absolute digimatic IP66 of Mitutoyo) of the punched out electrodes.

6.3 Computational Methods

The approach applied in this chapter is illustrated in Fig. 6.1. Related to this, the empiric surrogate models are introduced.

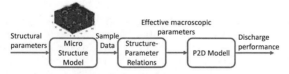

Figure 6.1: Flow chart of the surrogate model approach: Sample structure generation/evaluation, ESMs and electrochemical simulations in P2D.

6.3.1 Micro Structure Generation and Evaluation

The algorithms for the structure generation and evaluation are introduced in Chapter 5 for a lithium-based all-solid-state battery. Here, this approach is modified to account for liquid electrolyte. General assumptions for the generated micro structure are as follows:

---

[1] Cell assembly and execution of electrochemical tests were executed by the student assistant Christopher Hirsch. The author gratefully acknowledge his efforts in the laboratory.

homogeneous bulk within each phase, isotropic properties, no structure changes along layer thickness, no breakage of active material particles and particles of the Carbon Black-binder matrix (CBM), and no voids in the electrode.

The structure generation starts with setting randomly distributed nuclei with points $\zeta_{AM,i}$ in entire $\mathcal{M}$ for the active material phase AM. A nucleus fills the entire voxel. To every nucleus a random numerical particle size $S_{AM,i} \in \mathbb{N}$ is assigned. The geometrical particle radius R_p can be derived from S by $R_p = S \cdot \Delta x$. Afterwards, nuclei for the CBM phase at $\zeta_{CBM,i}$ are set. In contrast to AM nuclei, the nuclei at $\zeta_{CBM,i}$ are set in a sub-domain of $\mathcal{M}$ which is

$$\Omega_{CBM}^* = \{\varphi \in \mathcal{M} \mid \|\varphi - \zeta_{AM,i}\| > S_{AM,i} \; \forall i \wedge \mathcal{N}(\varphi) > 0\}, \tag{6.5}$$

wherein $\mathcal{N} \in \mathbb{N}^3$ is of the same size as $\mathcal{M}$ and contains the number of neighboring voxels containing active material for every voxel φ. Basically, this means that CB is only placed at the surface of the active material, which reproducers the experimentally observed structure of electrodes. To every $\zeta_{CBM,i}$ a random numerical particle size $S_{CB,i}$ is assigned. This kind of distribution was only one case in the previous investigation of different ASSB electrodes. Here, this distribution is the only option as a liquid electrolyte is used. The second change in comparison to Chapter 5 is the second solid species beside AM is the Carbon Black-binder matrix and not solely Carbon Black. The following evaluation of the generated structures is exactly as described in Chapter 5.

6.3.2 Electrochemical Modeling

The applied electrochemical model based on the work of Doyle et al. [54] and Legrand et al. [146] is introduced in Section 2.3.2. This model is extended with the enhanced structure-model parameter relations derived in Section 6.4.2.

As the whole calendering influence shall be simulated with one parameter set, the extended model applies empirical surrogate models (ESM) parameterized from 3D micro structure simulation, see Eqs. 6.10b, 6.12 and 6.14, while the classical P2D model uses the structure-parameter relations of Eqs. 6.1, 6.3 and 6.4. All other governing equations remain unchanged.

In Chapter 3, the parameters of the classical P2D model are estimated. Parts of the parameter estimation are repeated in this chapter, as additional experimental data are available due to the different calendering rates. Also, due to the ESMs the model has changed slightly. Parameter are estimated applying a least-square algorithm to minimize the deviation of simulated and experimental discharge curves from 0.5C to 5C. See Equation 6.6. As a three-electrode setup was used in the experiments, half-cell potentials of both electrodes are used in the target function.

For both models, exchange current densities, $i_{0,k} \forall k \in \{a, \; c\}$, and solid phase diffusion coefficients $D_{s,k} \forall k \in \{a, \; c\}$, of both electrodes are derived from the experiments, as well as the diffusion coefficient of the electrolyte D_e. For the parameter estimation, values of $D_{s,k}$ etc. taken from Chapter 3 are used as starting values. For the extended model, the bulk conductivity of NMC is derived from the experiments, while the conductivity of the Carbon Black-binder matrix is kept constant at $\bar{\sigma}_{c,CBM} = 760 \, \mathrm{S \, m^{-1}}$[76]. For the classical model the solid phase bulk conductivity of the cathode is adjusted which does not distinguish between the contribution of active material and Carbon Black-binder matrix. The anode conductivity is not adjusted since in literature a high conductivity of graphite electrodes is stated, e.g. $100\,\mathrm{S\,m^{-1}}$ in Ref. [98]. Also Chapter 3 of this work proofed it to be insensitively high.

The least square formulation for the parameter estimation is as follows

$$
\epsilon(\Theta) = \sum_j \sum_k \sum_i \left(\left(\frac{U_{\mathrm{cell,sim},j,k}(\Theta, t_i) - U_{\mathrm{cell,exp},j,k}(t_i)}{U_{\mathrm{ref}}} \right)^2 \right.
$$
$$
\left. + \left(\frac{\phi_{\mathrm{c,sim},j,k}(\Theta, t_i) - \phi_{\mathrm{c,exp},j,k}(t_i)}{U_{\mathrm{ref}}} \right)^2 \right), \tag{6.6}
$$

wherein Θ is the parameter set and ϵ the cost function to minimize. The simulated cell voltage is $U_{\mathrm{cell,sim},j,k}(\Theta, t_i)$, and $\phi_{\mathrm{c,sim},j,k}(\Theta, t_i)$ is the simulated cathodic half cell potential at the C-rate with index k and calendering rate with index j at the equidistant sample time t_i. In extend to the least square formulation in Chapter 3, Eq. 6.6 also considers different calendering ratios.

The parameter estimation routine leads to six adjustable parameters, which is comparable to published models, e.g. see Refs. [115, 73]. Lenze et al. highlighted the different distinguishable influences of different parameters in C-rate tests [73]. Further, Eq. 6.6 considers 4 C-rates and 6 calendering rates simultaneously.

6.4 Results and Discussion

In this section, micro structure simulations are presented and empiric surrogate models for the structure-model parameter relations are derived from 3D-simulations. Eventually, electrochemical simulation results are shown and validated with the experimental results of the calendering study to prove the concept of the enhanced structure-model parameter from the micro structure model.

6.4.1 Micro Structure Simulations

In this section, different micro structures are investigated for a variation of porosity. All simulations are run with a constant volume ratio of $1/4.4$ between the Carbon Black-binder Matrix and active material which is in accordance to the experimental cells in Section 6.2. Reference mean particle sizes of AM and CBM are $5.5\,\mathrm{\mu m}$ and $1.83\,\mathrm{\mu m}$, respectively. The voxel size is $0.33\,\mathrm{\mu m}$. The CBM particle size represents agglomerates containing CB and binder. Related to Refs. [27, 76, 173], conductivities are assumed as follows: $\tilde{\sigma}_{\mathrm{AM}} = 1.4 \times 10^{-2}\,\mathrm{S\,m^{-1}}$, $\tilde{\sigma}_{\mathrm{CB}} = 100\,\mathrm{S\,m^{-1}}$ and $\tilde{\sigma}_{\mathrm{ion}} = 0.6\,\mathrm{S\,m^{-1}}$. These values, are used to generate the samples for identification of the surrogate models. Literature values are used here, to ensure that the sample conductivity fit the magnitude of the experimental conductivity. The exact conductivity of the materials are determined by parameter estimation as described in Section 6.3.2. Literature values can only provide a initial guess, as differences between different probes of the same material are large. For instance, for NMC conductivities are reported over three magnitudes [97, 98, 76].

In Fig. 6.2, the interface area between active material and electrolyte is shown for different porosities. As this work is related to calendering, all plots are plotted with decreasing porosity respectively increasing calendering rate from left to right. Calendering rate is initial (non-calendered) layer thickness divided by layer thickness of the calendered electrode. Blue dots represent the 3D simulations. The dashed line represents the

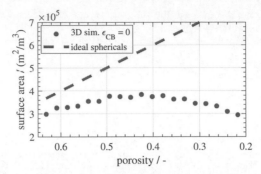

Figure 6.2: Effective surface area vs. electrode porosity. 3D simulations are dots. Eq. 6.4 is the dashed line. Volume ratio between AM and CB is 1/4.4.

widely used equation for active surface area, Eq. 6.4. At high porosities, surface area increases for the 3D structures as well as Eq. 6.4, as the number of particles increases and particle-to-particle contacts are marginal. The simulations show a maximal active surface area at a porosity of 0.45. At low porosities, the active surface area is decreased by particle-particle contacts. The influence of the electrode composition on electric and

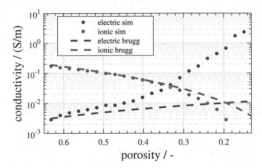

Figure 6.3: Effective electric (blue) and ionic (orange) conductivity vs. porosity, resulting from 3D simulations (dots) vs. Bruggeman relations (dashed lines). Constant active material-to-Carbon Black volume ratio of 1/4.4.

ionic conductivity is depicted in Fig. 6.3.

The simulated ionic conductivity (orange dots) decreases with decreasing porosity. The respective Bruggeman relation (orange dashed line) is in good accordance at high porosities. At low porosities, the simulated ionic conductivity decreases more significantly than the Bruggeman relation would assume, as first non-connected pores occur, i.e. pores inactive for the macroscopic charge transport. Thus, there is a minimal porosity above zero, at which effective conductivity drops to zero. Calendering should stay well above this porosity.

The simulated electric conductivity (blue dots) increases slightly towards higher porosities, until at a porosity of about 0.35 the percolation of the Carbon Black network starts and the electric conductivity increases more significantly. The Bruggeman relation for the solid phase conductivity (blue dashed line) shows a small deviation at high porosities. At moderate to low porosities, the difference is several orders of magnitude, as the influence of the Carbon Black on the conductivity, which is neglected in the Bruggeman relation, is large. Thus, there is a percolation threshold above which Carbon Black forms a network containing only Carbon Black. This enables a high conductivity of the entire electrode.

As electric and ionic conductivity are both relevant for the discharge performance of a lithium-ion battery, the intercept of electric and ionic conductivity in Fig. 6.3 could be a starting point to minimize the cell resistance and to maximize cell performance. Therefore, simulation results in Fig. 6.3 would suggest an optimal porosity of about 0.35. Applying the Bruggeman relation for electric and ionic conductivity would suggest an optimal porosity of 0.2 at a three times smaller conductivity. This highlights the importance of the usage of 3-dimensional models to obtain and understand effective conductivities, and to transfer this knowledge into electrochemical models.

6.4.2 Derived Algebraic Structure-Parameter Relations

Micro structure-model parameter relations applied in literature were reviewed in Section 6.1. In the following, based on structure simulations, three more accurate surrogates for those equations are derived: Equations 6.10b, 6.12 and 6.14. It should be noted that any type of surrogate model which fits the sample data could be used. Here, we try to stick to simple empiric equations which cover none-the-less essential physical effects as explained in the following.

For the further structure-model parameter relations, effective volume fractions are introduced as part of the empiric surrogates:

$$\varepsilon^*_{\text{AM}} = \frac{\varepsilon_{\text{AM}} - \varepsilon_{\text{crit,s}}}{1 - \varepsilon_{\text{crit,s}}}, \quad \forall \varepsilon_{\text{AM}} : \varepsilon_{\text{AM}} > \varepsilon_{\text{crit,s}}, \tag{6.7}$$

$$\varepsilon^* = \frac{\varepsilon - \varepsilon_{\text{crit,liq}}}{1 - \varepsilon_{\text{crit,liq}}}, \quad \forall \varepsilon : \varepsilon > \varepsilon_{\text{crit,liq}}, \tag{6.8}$$

$$\varepsilon^*_{\text{CBM}} = \varepsilon_{\text{CBM}} \frac{\varepsilon_{\text{CB}} + \varepsilon_{\text{AM}} - \varepsilon_{\text{crit,s}}}{1 - \varepsilon_{\text{crit,s}}}, \quad \forall \varepsilon_{\text{CBM}} : \varepsilon_{\text{CBM}} + \varepsilon_{\text{AM}} > \varepsilon_{\text{crit,s}} \tag{6.9}$$

For electrolyte and AM the effective volume fractions represent the relative distance between the percolation threshold volume fraction of the individual phase and a volume fraction of one. Values range between zero and one. This allows to combine the simplicity of the Bruggeman relation with the existence of the percolation threshold of the conducting network. The effective volume fractions of active material $\varepsilon^*_{\text{AM}}$, as well as the effective porosity ε^*, consider a critical percolation threshold of the conducting phase $\varepsilon_{\text{crit}}$, respectively for the solid and liquid phase: $\varepsilon_{\text{crit,s}}$ and $\varepsilon_{\text{crit,liq}}$. The volume fraction $\varepsilon_{\text{crit,s}}$ is related to the combined volume of AM and CBM. For the effective Carbon Black-binder volume fraction $\varepsilon^*_{\text{CBM}}$, the strong interaction with the active material is additionally taken into account. Eq. 6.9 is equal ε_{CBM} for $\varepsilon_{CBM} = 1 - \varepsilon_{AM}$ and zero for $\varepsilon_{CBM} + \varepsilon_{AM} = \varepsilon_{crit,s}$. Thus, the effective volume fraction scales linearly between zero, at $\varepsilon_{crit,s}$, and the absolute volume fraction ε_{CBM}, where all voxels of one phase are connected. The difference between effective volume fraction of CBM and Eqs. 6.7 and 6.8 gives weight to the assumption that in a dense AM structure the CBM percolates at lower volume fractions than in a less dense AM structure. This can be expected as dense structures have lower active material surface areas [175, 6], and thus, the CBM spreads to a smaller area. This leads to a higher surface-area specific coverage of the AM with CBM and thus to a lower percolation threshold.

For the ion conducting phase, the Bruggeman relation is extended regarding the effects of percolation and Carbon Black. In addition, a nonzero critical percolation porosity $\varepsilon_{crit,liq}$ is considered in ε_{CBM}^* (see Eqs. 6.8 and 6.10a) as well as an increase of tortuosity due to Carbon Black, represented by β_2 (see Eq. 6.10b):

$$\sigma_{liq} = \tilde{\sigma}_{liq} \cdot (\varepsilon^*)^{\beta_1}, \tag{6.10a}$$

$$\sigma_{liq} = \tilde{\sigma}_{liq} \cdot (\varepsilon^*)^{\beta_1 + \beta_2}. \tag{6.10b}$$

Eq. 6.10a, which is strongly related to Bruggeman, already provides a quite accurate fit, but for electrodes containing CBM Eq. 6.10b outperforms it due to the additional empiric term β_2 which is:

$$\beta_2 = \varepsilon_{CBM}^{\nu_1}. \tag{6.11}$$

Eq. 6.11 is related to an increase of liquid phase tortuosity due to Carbon Black, just as the Bruggeman coefficient for the active material phase. Due to that, Eq. 6.10b is used in the further. The fitting parameters of these equations are $\varepsilon_{crit,liq}$, an exponent β_1 which is related to the tortuosity due to the active material like a Bruggeman coefficient and a second coefficient ν_1.

For the electron conducting phase, Equation 6.12 is introduced, which consists of two summands. Due to the shape of the tangens hyperbolicus, the first summand is zero below the percolation threshold, is equal a Bruggeman type term, $\tilde{\sigma}_{CBM} \cdot (\varepsilon_{CBM}^*)^{\beta_3}$, representing the percolating CBM network above the percolation threshold, and has a steep transition between the both states. The mixed contribution of AM and CBM in the second summand can be understood as representation of a structure where CBM is present but does not form a network yet, but contribute to the AM network conductivity by e.g. connecting two AM particles. The mathematical formulation of the mixed contribution is related to a serial conduction of conducting AM element and a conducting AM element in parallel to a CBM element. This represents the physical processes in the electrode where the electron transport occurs partly in both materials, but the Carbon

Black-binder domain is not forming a matrix yet.

$$\sigma_s = \underbrace{\tilde{\sigma}_{CBM} \cdot (\varepsilon^*_{CBM})^{\beta_3} \cdot \frac{1}{2}\left(1 + \tanh\left(\frac{1}{\nu_2}\varepsilon^*_{CBM} - \nu_3\right)\right)}_{\text{contribution of percolated CB matrix}}$$

$$+ \left(\underbrace{\frac{1}{\tilde{\sigma}_{CBM} \cdot (\varepsilon^*_{CBM})^{\beta_3} + 2\tilde{\sigma}_{AM} \cdot (\varepsilon^*_{AM})^{\beta_4}}}_{\text{contribution of AM and non-percolated CBM}} + \underbrace{\frac{1}{2\tilde{\sigma}_{AM} \cdot (\varepsilon^*_{AM})^{\beta_4}}}_{\text{AM contribution}}\right)^{-1} . \qquad (6.12)$$

The exponents β_3 and β_4 are related to the tortuosity of the two conducting materials and the coefficients ν_2 and ν_3 are dependent on the percolation threshold and the slope in the transition.

The fitting parameters are $\varepsilon_{\text{crit,s}}$, β_3, β_4, ν_2, and ν_3. While Eq. 6.12 may seems quite long, for the absence of Carbon Black, it is equivalent to:

$$\sigma_s = \tilde{\sigma}_{AM} \cdot (\varepsilon^*_{AM})^{\beta_4} . \qquad (6.13)$$

For the effective surface area, Equation 6.14 is introduced:

$$a_S = \left(1 - \nu_4\frac{\varepsilon^{\nu_5}_{CBM}}{\varepsilon_{AM}}\right)\nu_6\frac{1 - 4\left(0.75 - \varepsilon_{AM}\right)^2 \cdot \varepsilon}{R_{p,AM}} . \qquad (6.14)$$

The fitting parameters are ν_4, ν_5 and ν_6. The first term is related to the blocking of active material surface by Carbon Black which is dependent on the ratio of the two volume fractions. The bigger the volume fraction of CBM compared to AM, the less surface is accessible for electrolyte. The numerator of the second term is a downward parabola vs. ε_{AM} which takes into account that there is an increasing particle-to-particle contact area as well as less electrolyte with increasing volume fraction of active material. The denominator is the particle size, representing the influence of the particle size on the surface-to-volume ratio of particles just as in Eq. 6.4.

To derive algebraic structure-model parameter relations, which in their entity are referred to as surrogate model, a set of sample structures is simulated applying the micro structure model. The different samples have varying electrode compositions and electric conductivities of Carbon Black. An excerpt of this data was presented above.

Table 6.1: Parameter ranges of sample data for parameterization of the surrogate model.

parameter	ε_{AM}	$\varepsilon_{CBM}/\varepsilon_{AM}$ ratio	$\tilde{\sigma}_{CBM}/\tilde{\sigma}_{AM}$ ratio
lower bound	0.4	0	4810
upper bound	0.75	0.3	14286

The ranges of the sample data are given in Tab. 6.1. Validity of surrogate model is only ensured for the sampled ranges. On the one hand, higher conductivity ratios can likely be calculated correctly as above a ratio of 14286 the physical process of percolation remains unchanged. On the other hand, the lower threshold is more relevant as below the bound, conduction through AM and conduction through CBM is not anymore as distinct as it is at high conductivity ratios. Exemplary fits of electric and ionic conductivity are shown in Figs. 6.4 and 6.5, respectively. They show percolation effects on the electric conductivity. The accordance for the ionic conductivity is higher than for the electric conductivity. This is related to the more complex processes for electrode conducting due to two conducting species compared to only on ion conducting species. Considering a smaller range of $\tilde{\sigma}_{\mathrm{CBM}}$-to-$\tilde{\sigma}_{\mathrm{AM}}$ ratios would increase the accuracy of the fit for an individual $\tilde{\sigma}_{\mathrm{CBM}}$-to-$\tilde{\sigma}_{\mathrm{AM}}$ ratio but would constrain the parameter range for the later parameter estimation.

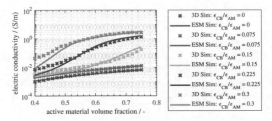

Figure 6.4: Validation of surrogate model with 3D simulation data. Effective electric conductivity vs. active material volume fraction for various CBM contents for the 3D-simulations (crosses) and the surrogate model (lines) for $\tilde{\sigma}_{\mathrm{CB}}$ of $67.33\,\mathrm{S\,m^{-1}}$ and $\tilde{\sigma}_{\mathrm{AM}}$ of $0.014\,\mathrm{S\,m^{-1}}$.

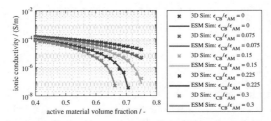

Figure 6.5: Validation of surrogate model with 3D simulation data. Effective ionic conductivity vs. active material volume fraction for the 3D-simulations (crosses) and the surrogate model (lines) with an intrinsic ionic conductivity of $5 \times 10^{-4}\,\mathrm{S\,m^{-1}}$

Parameter estimation with the sample data leads to the parameters listed in Tables 6.2, 6.3, and 6.4 for Eqs. 6.10b, 6.12, and 6.14, respectively.

Table 6.2: Empiric surrogate model parameters of Eq. 6.10b.

Parameter	$\varepsilon_{\mathrm{crit,liq}}$	β_1	ν_1
Value	0.127	1.77	0.680

Table 6.3: Empiric surrogate model parameters of Eq. 6.12.

Parameter	$\varepsilon_{\text{crit,s}}$	β_3	β_4	ν_1	ν_2
Value	0.1	0.023	2.0	0.2	1.5

Table 6.4: Empiric surrogate model parameters of Eq. 6.14.

Parameter	ν_4	ν_5	ν_6
Value	0.904	1.127	4.912

6.4.3 Coupled Electrochemical Simulations

In order to compare the electrochemical model with the derived new structure-model parameter relations to the classical model, both are parameterized using the experimental discharge curves from the calendering study. It should be noted that the adjusted parameters consider all calendering rates simultaneously, consequently there are no individual parameter sets for different calendering rates. The parameter sets are provided in Table 6.5. In contrast to the novel approach introduced in this work, calendering simulations, e.g. by Lenze et al., used one parameter set per calendering rate [73].

The discharge capacities for different calendering rates from 0.2C to 5C and for varying porosity are depicted for the classical model in Fig. 6.6. Squares represent experimental

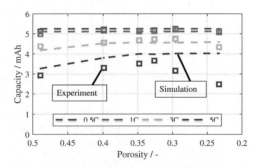

Figure 6.6: Discharge capacity at 2.9 V for different C-rates in a calendering study with classical model (dashed lines) vs. experiment (squares) with a constant Carbon Black-to-active material volume ratio of 1/4.4.

data, dashed lines are simulations. Each column of squares in this plot represents one cell. From 0.5C to 1C, the experimental and simulated discharge capacities show no impact of the calendering rate. This is because, the discharge capacity is close to the theoretical capacity and is limited solely by solid phase diffusion. At 5C and high porosities the simulation reproduces the experiment, but at low porosities the simulated discharge capacity does not decrease, while the experimental capacity does. Hence, the classical model fundamentally fails to simulate the transition from an electrode limited due to poor electrical conductivity at high porosities to an electrode limited by poor ionic conductivity at low porosities.

In Fig. 6.7, the same experimental data are plotted together with the simulation results of the extended model. At 0.5C and 1C also this model shows no influence of porosity

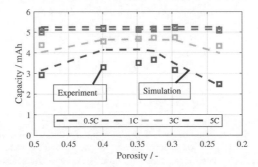

Figure 6.7: Discharge capacity at 2.9 V for different C-rates in a calendering study with extended model (dashed lines) vs. experiment (squares) with a constant active material-to-Carbon Black volume ratio of 1/4.4.

on discharge capacity. For experiment and simulation, at 3C there is a slight and for 5C there is a distinct optimum of the discharge capacity at a porosity of about 0.35. The simulations reproduce experimental trends qualitatively nicely, though the capacity at 5C and 40 % porosity is too high. Still, the experimental trend is correctly reproduced with the proposed model. Thus in contrast to the classical model, the extended model is feasible to simulate the transition from an electric conduction limited electrode to an ionic conduction limited electrode with increasing calendering rate. Residuals of the parameter estimation of both models are depicted in Fig. 6.11 for quantitative comparison. The normalized residual of the extended model is about 9 % lower than for the classical model.

In Figs. 6.8 to 6.10, discharge curves for the experiments and the extended model are shown for four C-rates and 3 different calendering rate. For all C-rates and calendering rates simulation and experiment are in good accordance, as well with respect to cell voltage, as with respect to half-cell potential of the cathode. Thus, the extended model is feasible to predict the discharge performance as well as the overpotentials of the cell.

In conclusion, the extended model reproduces the calendering experiment while the classical model fails. From this, it is concluded that calendering effects are strongly related to change of porosity and the related conductivities and active surface area considered by the extended model. Further effects may occur due to surface resistance etc.

6.4.4 Evaluation of the Robustness of the Approach

In Section 6.4.3, all experimental data are used for the parameter estimation of the model. If the model shall be used for optimization of the calendering process, it has to be able to reproduce the experiment without using all presently available calendering data sets. To investigate the feasibility of the electrochemical model with the enhanced structure-model parameter relations to predict the calendering impact properly, a scenario is

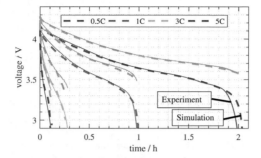

Figure 6.8: Model validation with experimental data from a C-rate test in a three-electrode setup. Uncalendered cathode with a porosity of about 48.9 % and an active material-to-Carbon Black-binder volume ratio of 1/4.4. Simulated (dashed lines) and experimental (solid lines) discharge curves. Gray lines represent the simulated (dashed lines) and experimental (solid lines) half-cell potential of the cathode. The extended model is used for simulations.

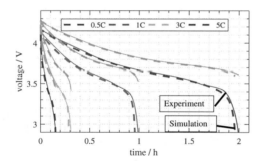

Figure 6.9: Model validation with experimental data from a C-rate test in a three-electrode setup. Moderately calendered cathode with a porosity of about 32.6 % and an active material-to-Carbon Black-binder volume ratio of 1/4.4. Simulated (dashed lines) and experimental (solid lines) discharge curves. Gray lines represent the simulated (dashed lines) and experimental (solid lines) half-cell potential of the cathode. The extended model is used for simulations.

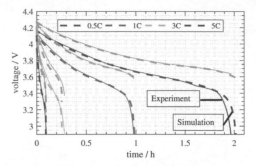

Figure 6.10: Model validation with experimental data from a C-rate test in a three-electrode setup. Highly calendered cathode with a porosity of about 23.3 % and an active material-to-Carbon Black-binder volume ratio of 1/4.4. Simulated (dashed lines) and experimental (solid lines) discharge curves. Gray lines represent the simulated (dashed lines) and experimental (solid lines) half-cell potential of the cathode. The extended model is used for simulations.

assumed wherein only the non-calendered cell and one single calendering rate is available. For this purpose, the parameter estimation and C-rate test simulations from Section 6.4.3 are repeated for the extended model with the subset of data which consists of the non-calendered cathode with a porosity of 48.9 % and the cathode calendered to a layer thickness of 62 µm with a porosity of 32.6 %, respectively uncalendered and highly calendered cell (23.3 % porosity). The estimated parameters are listed in Table

Table 6.5: Estimated parameters of the electrochemical model for different scenarios and models. $\bar{\sigma}_{c,CBM} = 760\,\mathrm{S\,m^{-1}}$ [76].

Model	extended	classical	extended	extended
evaluated cells	all cells	all cells	cells 1 & 4	cells 1 & 6
parameter				
$D_{s,c}$ in $\mathrm{m^2\,s^{-1}}$	4.976×10^{-12}	5.000×10^{-12}	4.976×10^{-12}	4.512×10^{-12}
$D_{s,a}$ in $\mathrm{m^2\,s^{-1}}$	1.177×10^{-14}	11.097×10^{-14}	1.180×10^{-14}	0.879×10^{-14}
D_e in $\mathrm{m^2\,s^{-1}}$	5.927×10^{-10}	6.693×10^{-10}	5.931×10^{-10}	6.058×10^{-10}
$\bar{\sigma}_{c,AM}$ in $\mathrm{S\,m^{-1}}$	0.0161	-	0.0161	0.0167
$\sigma_{s,AM+CBM}$ in $\mathrm{S\,m^{-1}}$	-	0.0141	-	-

6.5 in comparison to those using the full data set and those using the classical model. Respective residuals are shown in Fig. 6.11. Consideration of the uncalendered and a moderately calendered cell leads to almost exact quantitatively the same parameters.

Thus, even conducting only two experiments allow parameterization of the extended model and prediction of the optimal porosity with a precision of 5 percentage points of porosity (see Fig. 6.7). Further experimental tests would be required to find the optimal calendering rate in this scenario as 5 percentage points of porosity is not small compared to the parameter space. However, the model could reduce the total costs by reducing

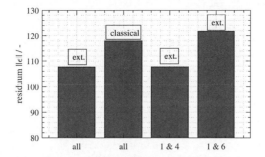

Figure 6.11: Residuals of the extended model (ext.) considering all cells for parameter estimation (PE), classical model considering all cells for PE, extended model considering an uncalendered and a moderately calendered cell (cells 1 & 4), and extended model considering an uncalendered and a highly calendered cell (cells 1 & 6). The residual is defined in Eq. 6.6 and is the sum of the squared deviation of cell voltage and cathode half-cell potential for all C-rates and all calendering rate, respectively, with 100 equidistant sample points per curve.

the number of required experiments in an combined experimental and simulation-based optimization process compared to a solely experimental optimization.

There are different possible reasons of the discrepancy between measured and simulated discharge performance. In general, influences of the micro structure model, the surrogate model, the homogeneous model, as well as deviations in the experimental data are possible. The micro structure model provides an uncertain predictions of the effective parameters, as there is a discretization error, as well as there is a possible discrepancy between the real micro structure and the artificial structure which is evaluated. Lastly, the experimental data is based on cells with a diameter of 18 mm. Due to that, there could be an effect of manufacturing deviations on the cell properties. Electrodes were manufactured in a large-scale role-to-role process. The length-scale of process deviations are bigger than the cell size. The leads to possible cell-to-cell deviations.

Eventually, the estimated parameter set has to be unique, proving physical insight. The estimated parameters are in the rage of quantitative values provided in literature. Chen et al. reported a NMC bulk conductivity of $1.06 \times 10^{-3} \, \mathrm{S \, m^{-1}}$ and electrical conductivity of the Carbon Black-binder matrix of $760 \, \mathrm{S \, m^{-1}}$. For a slightly different electrolyte than applied here they stated a diffusion coefficient of $1.2 \times 10^{-9} \, \mathrm{m^2 \, s^{-1}}$ [76]. Vazquez-Arenas et al. stated diffusion coefficients for graphite and NMC of $3.9 \times 10^{-14} \, \mathrm{m^2 \, s^{-1}}$ and $1.64 \times 10^{-14} \, \mathrm{m^2 \, s^{-1}}$, respectively. The estimated parameters are in good accordance with the literature, while deviations are plausible since slightly different materials are used.

6.5 Concluding Remarks

Accurate structure-model parameter relations are essential for prediction and optimization of cell properties. This is especially important during electrode manufacturing, e.g. for the calendering process. Therefore, a 3D micro structure model was applied to derive more accurate structure-model parameter relations than those applied in the classical Doyle-Newman model. These include addition of Carbon Black and particle sizes and distributions. Artificial non-spherical particles are generated in the micro structure model and the effective electric conductivity of an electrode, its ionic conductivity and interface area between active material and electrolyte are determined for various electrode compositions. Empirical surrogate models were derived from the conductivity relations of the micro structure model and are used to extend the classical Doyle-Newman model. The such extended electrochemical model is able to reproduce and predict the calendering experiment. A single experimental calendering rate allowed to estimate the optimal calendering rate with an accuracy of about 5 percentage points. The feasibility of the extended model to reproduce the experiment suggests that homogeneous models could be sophisticated enough to simulate the micro structure and calendering influence on the cell performance for the investigated system. It is concluded that the investigated calendering rate primarily effects the porosity of the electrode. Consideration of effects like CB particle breaking or interface resistances were not required to predict the cell performance in dependence on the calendering rate.

Chapter 7

Summary and Conclusions

This dissertation has addressed challenges and applications of electrochemical modeling in the context of lithium-based batteries to assess the feasibility of mathematical modeling to enable knowledge-based design of battery production. Therefore, the required model complexity of electrochemical models was investigated. To this end, an electrochemical pseudo-2D battery model was parameterized and the uniqueness of the parameter set was ensured applying a combination of multi-start/multi-step parameter estimation and using experimental data of an open cell voltage measurement, C-rate tests, and electrochemical impedance spectroscopy. All measurements were conducted in a three-electrode setup, allowing to measure half-cell potentials. It was shown, that this experimental effort was required to get a unique parameter set as even with the half-cell potentials, the P2D model was not identifiable from the C-rate test. These results lead to the conclusion, that any modeling efforts should be supported by OCV, C-rate, and EIS measurements. Indeed, EIS measurements could be replaced by offline diagnosis, e.g. to measure the electric electrode conductivity of a dry electrode.

Based on the parameterized P2D model, uncertainty quantification was carried out, to investigate the influence of electrode properties, affected by the entire electrode production, on the cell performance. The results, revealed the nonlinearity of the P2D model, causing issues with the applied nested point estimate method for the global sensitivity analysis. To validate the point estimate method results, Monte Carlo simulations were conducted. It was shown, that the sensitivity was strongly affected by the applied C-rate and the gradient of the OCV curve of the respective electrode. Thus, the model-based uncertainty quantification has to be repeated for any battery in any specific application. If modeling and simulations are not considered in the design of the production, this can lead to an inefficient cell balancing and too high or too low quality bounds, causing increased cost or increased rejection rate.

While the uncertainty quantification is parameter-based and thus assesses the entire production, a 3D micro structure model was introduced to investigate the influence of the mixing routine of an all-solid-state battery on electric and ionic conductivity and the influence of the calendering step on the cathode of a lithium-ion battery. Simulation of different ASSB electrode compositions and mixing protocols revealed the optimal conducting additive volume fraction to ensure a percolation of the conducting network of the additive, as well a sufficient ionic conductivity. Different mixing routines showed contrary effects on either electric or ionic conductivity. Thus, optimization of the mixing would require coupling of the micro structure model with an electrochemical model as was done in this work to investigate the calendering influence on the cell performance. Therefore, from the micro structure model, empiric surrogate models were derived, providing enhanced structure-model parameter relations between electrode composition and model parameters such as effective conductivity. These enabled the P2D model to simulate the transition from an electrically limited electrode at low calendering rates to an ionically limited electrode at high calendering rates with a single parameter set.

The need for only one parameter set, in contrast to previous work on the calendering influence as Ref. [73], enables the model-based optimization of the calendering rate and thus an increased benefit of modeling and simulations for the battery design and production. Further, the model extension proofed that homogeneous battery models are able to simulate the battery performance with good accuracy and that the influence of small-scale local effects is marginal.

Concluding the thesis with respect to the hypothesis and scientific question in Chapter 1, it is highlighted that 3D-micro structure modeling is feasible to simulate the influence of electrode composition and mixing protocol on electrical and ionical conductivity, and that the coupling of the stationary 3D model with a dynamic P2D model is sufficient to model the calendering influence on the electrochemical model. The benefit of the coupling compared to a stand-alone P2D model is a wide range of validity for different electrode compositions, while the classical P2D model is still applicable for local investigations of a cell with parameters similar to the reference parameter set the model is validated for, e.g. an uncertainty quantification.

This thesis highlighted the potential but also the limitations of modeling in the context of production. In future research, matching the model sensitivities and a model validation for a parameter range is required to provide a significant impact to the design of enhanced LIBs and the development of next-generation batteries. Therefore, the approach applied in this work about coupling of micro structure models and electrochemical models should be continued. An enhancement of structure generation could lead to more accurate micro structures and hence more accurate structure-model parameter relations. Also, further parameters and further production steps should be addressed.

Lately, model identifiability has gained attention. In future research, this development should continue and should not only be addressed by groups with a theoretical perspective, but by the majority of groups applying models. This would probably require standardization of sample-based identifiability tests which are applicable for P2D models. This further would require making available of the programs, for instance as an open-access toolbox as Marelli et al. did with UQLab for uncertainty quantification [177]. Such an identifiability toolbox would allow an easy assessment of the accuracy of model parameters and would strengthen applicability of models and trust in model-based studies, e.g. in the context of lithium-based battery production.

Appendix A

Structural Identifiability of Parameter Estimation Steps 1 and 3 [1]

In Chapter 3.2.1, parameter estimation steps 1 (static parameters and OCV curve) and 3 (dynamic parameters and EIS) have not been investigated through multi-start parameter estimation. Only PE step 2 was assessed in-depth. The underlying reason is illustrated in the following.

Structural Identifiability of Parameter Estimation Step 1   PE step 1 has four parameters, $c_{c,0}$, $c_{a,0}$, $\Delta c_{c,max}$, $\Delta c_{a,max}$, and one experiment, half-cell open cell voltages, shown in Fig. A.1.

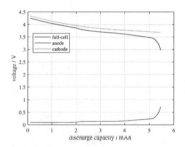

Figure A.1: Half-cell potentials of the OCV curve of the NMC vs. graphite lithium ion cell.

Features of the anode OCP are a wide flat range with discrete steps and a steep section at high discharge capacities (low SOCs). In contrast, the cathode OCP is steadily decreasing with increasing discharge capacity. As the OCP was measured in a three-electrode setup both electrodes share the abscissa of full-cell SOC. But, as this is neither a state variable nor a parameter of the model, $c_{c,0}$, $c_{a,0}$, $\Delta c_{c,max}$, and $\Delta c_{a,max}$ have to be estimated. As OCVs are formulated as functions vs. intercalation ratio, respectively normalized concentration $c_s/\Delta c_{max}$, Δc_{max} can be used to stretch the OCV of one electrode compared to full-cell SOC. The initial concentration $c_{s,0}$ of one electrode can be used to shift one electrode vs. full-cell SOC. This is possible for both electrodes. However, due to the unique features of the half-cell OCVs, e.g. the intercalation steps of the anode, you can not shift and stretch one electrode to reproduce the features of

---

[1]This chapter was first published by Laue and Krewer [95].

the second electrode. For an in-depth discussion of full-cell OCV reconstruction from half-cell OCVs it is referred to Refs. [178] and [24], wherein this is done for LIBs at different states of health.

Structural Identifiability of Parameter Estimation Step 3 The objective of PE step 3 is to distinguish between cathode solid phase conductivity $\sigma_{s,c}$ and cathode exchange current density $j_{0,c}$. To illustrate the feasibility of EIS to do this, parameter variations of $\sigma_{s,c}$ and $j_{0,c}$ are shown in Figures A.2 and A.3, respectively.

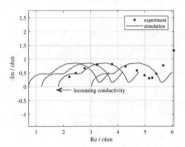

Figure A.2: Impedance spectrum of the OCV curve of the NMC vs. graphite lithium ion cell. Cathode solid phase conductivity is varied.

With increasing solid phase conductivity the semi-circles are moving to lower real parts of impedance, while the size of semi-circles stay the same. If the exchange current density is increased, the second semi-circle, related to the cathode, is decreasing in size, while the abscissa intercept at high frequencies remain the same. Thus, with position and size of the cathode semi-circle the two parameters $\sigma_{s,c}$ and $j_{0,c}$ can be distinguished. It is noteworthy, that the abscissa intercept is also dependent on the electrolyte conductivity. However, this is not adjustable parameter of the applied model as the electrolyte diffusivity is fitted in PE step 2 and the electrolyte conductivity is derived therefrom by the Nernst-Einstein equation. See Equation 2.26 in Section 2.3.2.

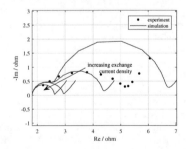

Figure A.3: Impedance spectrum of the OCV curve of the NMC vs. graphite lithium ion cell. Cathode exchange current density is varied.

Further, it is referred to studies on the structural identifiability of a linearized SPM from EIS data of Bizeray et al. They concluded that beside the two solid phase diffusion coefficient a lumped charge-transfer resistance could be identified [113]. However, it has to be noticed that their studies were based on full-cell data only. Thus, if half-cell potentials are available, charge transfer resistances of both electrodes are identifiable.

Appendix B

Comprehensive Results of Multi-Start Parameter Estimation [1]

Multi-start parameter estimation is conducted for PE step 2 as described in Section 3.1.3. Out of 152 initial points, 49 have been chosen by the algorithm due to the rejection

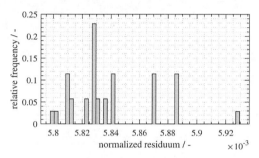

Figure B.1: Histogram of the normalized residuum F_2 of the entity of parameter estimations which fulfilled the initial requirement of 50 % of the experimental capacity at 0.5C.

criterion. Further, sets with a normalized residuum below 6.0×10^{-3} were chosen for the further analysis. Due to that, the number of used parameter sets is 35. Figure B.1 shows the normalized residuum of the parameter estimation defined in Equation 3.2. There is a narrow band of parameter sets with a residuum between 5.8×10^{-3} and 6.0×10^{-3}. Thus, they only vary about 3.3 % and all are equivalent to an excellent accordance between simulation and experiment. The rejected few starting points lead to a final residuum up to 14 % higher than gained from the optimal parameter set and were removed from the further analysis as stated before. Thus, the least-square solver is able to converge for a large number of initial points.

In Figure B.2, estimated parameters of the anode are shown. The solid phase diffusion coefficient was estimated to $6.76 \times 10^{-15} \, \mathrm{m^2 \, s^{-1}}$ with a standard deviation of about 7.00 %. This strongly indicates that this parameter is identifiable from the experiment. Anode solid phase conductivity varies from $0.58 \, \mathrm{S \, m^{-1}}$ up to $650 \, \mathrm{S \, m^{-1}}$ with about 60 % of values below $50 \, \mathrm{S \, m^{-1}}$. This result can be explained either by insensitivity with respect to this parameter or non-unique parameter sets. Anode exchange current density varies from $0.373 \, \mathrm{A \, m^{-2}}$ to $0.468 \, \mathrm{A \, m^{-2}}$ with a standard deviation of about 6.8 %. Noteworthy

---

[1]Parts of this chapter was first published by Laue and Krewer [95].

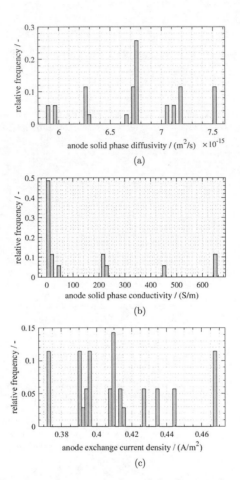

Figure B.2: Histogram of estimated values of the three anode parameters. (a) Anode solid phase diffusion coefficient, (b) anode solid phase conductivity, and (c) anode exchange current density.

is, that there is no clear peak at the average exchange current density. However, there is no indication of unidentifiability. In summary, for the anode, parameter estimation step 2 leads to a unique value for solid phase diffusion and exchange current density with an uncertainty of about 7 %. The solid phase conductivity is unidentifiable, probably due to insensitivity. This could be explain by the high conductivity of graphite-based anodes in general. The conductivity is so high that is does not lead to a significant potential loss, which could be exploited for parameter estimation.

In Figure B.3, estimated parameters of the cathode from the 35 least-square optimizations are shown. The cathode solid phase diffusion coefficient varies from $3.70 \times 10^{-13}\,\mathrm{m^2\,s^{-1}}$ to $7.73 \times 10^{-11}\,\mathrm{m^2\,s^{-1}}$. Compared to the anode, this deviation is huge. A large portion of parameter sets are at low diffusion coefficients. But, the inset plot in Figure B.3a with a logarithmic scale shows that at low diffusion coefficients, there is a broad distribution as well. In contrast to the anode diffusion coefficient, for the cathode this indicates unidentifiability of the cathode diffusion coefficient in PE step 2. In Figure B.3b, the cathode solid phase conductivity is shown. The existence of the two peaks at high conductivities shows that there are parameter sets in which $\sigma_{\mathrm{s,c}}$ is not sensitive. Besides few high conductivities, there is a relative small distribution of conductivities around $0.21\,\mathrm{S\,m^{-1}}$. This is in accordance to the results of the L-shaped plot in Figure 3.2. There is one arc, where the conductivity is locally identifiable at low conductivity and there is the second arc, where cathode solid phase conductivity is not sensitive. Last, the distribution of the cathode exchange current density is shown. Values vary around $1.30\,\mathrm{A\,m^{-2}}$ with a standard deviation of about 10.6 %. Summarizing parameter estimation of the cathode, diffusion coefficient and solid phase conductivity vary over several orders of magnitude but both have a distinct lower bound. This suggests, non-uniqueness as there are parameter sets with low values where the parameter is sensitive, and there are parameter sets where the parameter is not sensitive, which implies that there is another loss process which is sensitive instead.

It is likely that the position of peaks in the histograms is biased by the sample points given by Eq. 3.4, as well as by the settings of the least-square algorithm. However, to reject the hypothesis of uniqueness, the exact peak position is of minor interest.

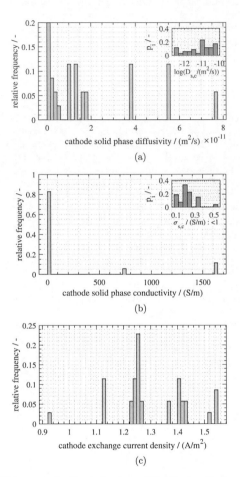

Figure B.3: Histogram of estimated values of the three cathode parameters. The y-axis is the relative frequency p_i. (a) Cathode solid phase diffusion coefficient, (b) cathode solid phase conductivity, and (c) cathode exchange current density. In (a), the inset shows the same data on a different scale. In (b), the inset shows all data smaller than $1\,\mathrm{S}\,\mathrm{m}^{-1}$.

Bibliography

[1] K. Brandt. Historical development of secondary lithium batteries. Solid State Ionics, 69(3-4):173–183, 1994.

[2] A. Jossen and W. Weydanz. Moderne Akkumulatoren richtig einsetzen. Ubooks Verlag, 1 edition, 2006.

[3] Q. Li, J. Chen, L. Fan, X. Kong, and Y. Lu. Progress in electrolytes for rechargeable Li-based batteries and beyond. Green Energy Environ., 1(1):18–42, 2016.

[4] Nationale Plattform Elektromobilität (NPE). Roadmap for an Integrated Cell and Battery Production in Germany. page 68, 2016.

[5] D. L. Wood, J. Li, and C. Daniel. Prospects for reducing the processing cost of lithium ion batteries. J. Power Sources, 275:234–242, 2015.

[6] V. Laue, N. Wolff, F. Röder, and U. Krewer. Modeling the Influence of Mixing Strategies on Micro Structural Properties of All-Solid State Electrodes. Energy Technol., 2019.

[7] M. S. Whittingham. Lithium Batteries and Cathode Materials. Chem. Rev., 104(607):4271, 2004.

[8] A. Mishra, A. Mehta, S. Basu, S. J. Malode, N. P. Shetti, S. S. Shukla, M. N. Nadagouda, and T. M. Aminabhavi. Electrode materials for lithium-ion batteries. Mater. Sci. Energy Technol., 1(2):182–187, December 2018.

[9] Y. Ding, R: Wang, L. Wang, K. Cheng, Z. Zhao, D. Mu, and B. Wu. A Short Review on Layered LiNi0.8Co0.1Mn0.1O2 Positive Electrode Material for Lithium-ion Batteries. Energy Procedia, 105:2941–2952, 2017.

[10] A. Mukanova, A. Jetybayeva, S. T. Myung, S. S. Kim, and Z. Bakenov. A mini-review on the development of Si-based thin film anodes for Li-ion batteries. Mater. Today Energy, 9:49–66, 2018.

[11] S. Goriparti, E. Miele, F. De Angelis, E. Di Fabrizio, R. Proietti Zaccaria, and C. Capiglia. Review on recent progress of nanostructured anode materials for Li-ion batteries. J. Power Sources, 257:421–443, 2014.

[12] J. G. Kim, B. Son, S. Mukherjee, N. Schuppert, A. Bates, O. Kwon, M J. Choi, H. Y. Chung, and S. Park. A review of lithium and non-lithium based solid state batteries. J. Power Sources, 282:299–322, 2015.

[13] A. Manuel Stephan and K. S. Nahm. Review on composite polymer electrolytes for lithium batteries. Polymer (Guildf)., 47(16):5952–5964, 2006.

[14] M. Tatsumisago, M. Nagao, and A. Hayashi. Recent development of sulfide solid electrolytes and interfacial modification for all-solid-state rechargeable lithium batteries. J. Asian Ceram. Soc., 1(1):17–25, 2013.

[15] A. B. Aziz, T. J. Woo, M. F. Z. Kadir, and H. M. Ahmed. A conceptual review on polymer electrolytes and ion transport models. J. Sci. Adv. Mater. Devices, 3(1):1–17, 2018.

[16] K. Kerman, A. Luntz, V. Viswanathan, Y.-M. Chiang, and Z. Chen. Review—Practical Challenges Hindering the Development of Solid State Li Ion Batteries. J. Electrochem. Soc., 164(7):A1731–A1744, 2017.

[17] F. Zheng, M. Kotobuki, S. Song, M. O. Lai, and L. Lu. Review on solid electrolytes for all-solid-state lithium-ion batteries. J. Power Sources, 389(April):198–213, 2018.

[18] P. Braun, C. Uhlmann, M. Weiss, A. Weber, and E. Ivers-Tiffée. Assessment of all-solid-state lithium-ion batteries. Journal of Power Sources, 393:119 – 127, 2018.

[19] K. M. Nairn, a. S. Best, P. J. Newman, D. R. MacFarlane, and M. Forsyth. Ceramic-polymer interface in composite electrolytes of lithium aluminum titanium phosphate and polyetherurethane polymer electrolyte. Solid State Ionics, 121(1):115–119, 1999.

[20] Y.-C. Jung, S.-M. Lee, J.-H. Choi, S. S. Jang, and D.-W. Kim. All Solid-State Lithium Batteries Assembled with Hybrid Solid Electrolytes. J. Electrochem. Soc., 162(4):A704–A710, 2015.

[21] J. Vetter, P. Novák, M. R. Wagner, C. Veit, K. C. Möller, J. O. Besenhard, M. Winter, M. Wohlfahrt-Mehrens, C. Vogler, and A. Hammouche. Ageing mechanisms in lithium-ion batteries. J. Power Sources, 147(1-2):269–281, 2005.

[22] J. Chakraborty, C. P. Please, A. Goriely, and S. J. Chapman. Combining mechanical and chemical effects in the deformation and failure of a cylindrical electrode particle in a Li-ion battery. Int. J. Solids Struct., 54:66–81, 2015.

[23] V. Agubra and J. Fergus. Lithium ion battery anode aging mechanisms. Materials, 6:1310–1325, 03 2013.

[24] B. Rumberg, K. Schwarzkopf, B. Epding, I. Stradtmann, and A. Kwade. Understanding the different aging trends of usable capacity and mobile li capacity in li-ion cells. Journal of Energy Storage, 22:336–344, 2019.

[25] L. Hoffmann, J.-K. Grathwol, W. Haselrieder, R. Leithoff, T. Jansen, K. Dilger, K. Dröder, A. Kwade, and K. Kurrat. Capacity distribution of large lithium-ion battery pouch cells in context with pilot production processes. Energy Technology, 2019.

[26] V. Laue, O. Schmidt, H. Dreger, X. Xie, F. Röder, R. Schenkendorf, A. Kwade, and U. Krewer. Model-based Uncertainty Quantification for the Product Properties of Lithium-Ion Batteries. Energy Technology, 2019.

[27] T. P. Heins, N. Schlüter, and U. Schröder. Electrode-Resolved Monitoring of the Ageing of Large-Scale Lithium-Ion Cells by using Electrochemical Impedance Spectroscopy. ChemElectroChem, 4(11):2921–2927, 2017.

[28] K. Smith and C.-Y. Wang. Power and thermal characterization of a lithium-ion battery pack for hybrid-electric vehicles. J. Power Sources, 160(1):662–673, 2006.

[29] H. Dreger, H. Bockholt, W. Haselrieder, and A. Kwade. Discontinuous and Continuous Processing of Low-Solvent Battery Slurries for Lithium Nickel Cobalt Manganese Oxide Electrodes. J. Electron. Mater., 44(11):4434–4443, 2015.

[30] W. Haselrieder, S. Ivanov, H.Y. Tran, S. Theil, L. Froböse, B. Westphal, M. Wohlfahrt-Mehrens, and a. Kwade. Influence of formulation method and related processes on structural, electrical and electrochemical properties of LMS/NCA-blend electrodes. Prog. Solid State Chem., 42(4):157–174, 2014.

[31] H. Graebe, A. Netz, S. Baesch, V. Haerdtner, and A. Kwade. A Solvent-Free Electrode Coating Technique for All Solid State Lithium Ion Batteries. ECS Transactions, 77:393–401, 2017.

[32] B.G. Westphal, H. Bockholt, T. Günther, W. Haselrieder, and a. Kwade. Influence of Convective Drying Parameters on Electrode Performance and Physical Electrode Properties. ECS Trans., 64(22):57–68, 2015.

[33] Z. Liu and P. P. Mukherjee. Microstructure Evolution in Lithium-Ion Battery Electrode Processing. J. Electrochem. Soc., 161(8):E3248–E3258, 2014.

[34] V. Laue, F. Röder, and U. Krewer. Joint structural and electrochemical modeling: Impact of porosity on lithium-ion battery performance. Electrochimica Acta, page S0013468619309065, May 2019.

[35] W. Haselrieder, S. Ivanov, H. Bockhold, and A. Kwade. Impact of the Calendering Process on the Interfacial Structure and the Related Electrochemical Performance of Secondary Lithium-Ion Batteries. ECS Transactions, 50:59–70, 2013.

[36] J. Shim and K. Striebel. Effect of electrode density on cycle performance and irreversible capacity loss for natural graphite anode in lithium-ion batteries. J. Power Sources, 119-121, 2003.

[37] G. F. Yang and S. K. Joo. Calendering effect on the electrochemical performances of the thick Li-ion battery electrodes using a three dimensional Ni alloy foam current collector. Electrochim. Acta, 170:263–268, 2015.

[38] A. van Bommel and R. Divigalpitiya. Effect of Calendering LiFePO4 Electrodes. J. Electrochem. Soc., 159(11):A1791–A1795, 2012.

[39] J. Smekens, R. Gopalakrishnan, N. Steen, N. Omar, O. Hegazy, A. Hubin, and J. Van Mierlo. Influence of Electrode Density on the Performance of Li-Ion Batteries: Experimental and Simulation Results. Energies, 9(2):104, 2016.

[40] D. Schmidt, M. Kamlah, and V. Knoblauch. Highly densified ncm-cathodes for high energy li-ion batteries: Microstructural evolution during densification and its influence on the performance of the electrodes. Journal of Energy Storage, 17:213 – 223, 2018.

[41] Z. Karkar, T. Jaouhari, A. Tranchot, D. Mazouzi, D. Guyomard, B. Lestriez, and L. Roué. How silicon electrodes can be calendered without altering their mechanical strength and cycle life. Journal of Power Sources, 371:136–147, 2017.

[42] Y. Sheng, C. R. Fell, Y. K. Son, B. M. Metz, J. Jiang, and B. C. Church. Effect of Calendering on Electrode Wettability in Lithium-Ion Batteries. Front. Energy Res., 2(December):1–8, 2014.

[43] T. Jansen, D. Blass, S. Hartwig, and K. Dilger. Processing of Advanced Battery Materials—Laser Cutting of Pure Lithium Metal Foils. Batteries, 4(3):37, 2018.

[44] J. Schmitt. Untersuchung zum Herstellungsprozess des Elektrode-Separator-Verbunds für Lithium-Ionen Batteriezellen. Vulkan Verlag, 2015.

[45] H. H. Heimes, C. Offermanns, A. Mohsseni, H. Laufen, U. Westerhogg, L. Hoffmann, P. Niehoff, M. Kurrat, M. Winter, and A. Kamper. The effects of mechanical and thermal loads during lithium-ion pouch cell formation and their impacts on process time. Batteries & Supercaps, 2019.

[46] S. J. An, J. Li, Z. Du, C. Daniel, and D. L. Wood III. Fast formation cycling for lithium ion batteries. Journal of Power Sources, 343:846–852, 2017.

[47] F. Röder, R. D. Braatz, and U. Krewer. Multi-Scale Simulation of Heterogeneous Surface Film Growth Mechanisms in Lithium-Ion Batteries. Journal of The Electrochemical Society, 164(11):E3335–E3344, 2017.

[48] F. Röder, R. D. Braatz, and U. Krewer. Multi-Scale Modeling of Solid Electrolyte Interface Formation in Lithium-Ion Batteries. In Computer Aided Chemical Engineering, volume 38, pages 157–162. 2016.

[49] F. Röder, V. Laue, and U. Krewer. Model Based Multiscale Analysis of Film Formation in Lithium-Ion Batteries. Batteries & Supercaps, February 2019.

[50] V. Ramadesigan, K. Chen, N. A. Burns, V. Boovaragavan, R. D. Braatz, and V. R. Subramanian. Parameter Estimation and Capacity Fade Analysis of Lithium-Ion Batteries Using Reformulated Models. J. Electrochem. Soc., 158(9):A1048, 2011.

[51] A. A. Franco. Multiscale modelling and numerical simulation of rechargeable lithium ion batteries: concepts, methods and challenges. RSC Advances, 3:13027–13058, 2013.

[52] B. Suthar, D. Sonawane, R. D. Braatz, and V. R. Subramanian. Optimal low temperature charging of lithium-ion batteries. IFAC-PapersOnLine, 28(8):1216–1221, 2015.

[53] M. Torchio, L. Magni, R. D. Braatz, and D. M. Raimondo. Optimal Health-aware Charging Protocol for Lithium-ion Batteries: A Fast Model Predictive Control Approach. IFAC-PapersOnLine, 49(7):827–832, 2016.

[54] M. Doyle. Modeling of Galvanostatic Charge and Discharge of the Lithium/Polymer/Insertion Cell. J. Electrochem. Soc., 140(6):1526, 1993.

[55] M. Doyle and J. Newman. The use of mathematical modeling in the design of lithium/polymer battery systems. Electrochim. Acta, 40(13-14):2191–2196, 1995.

[56] J. Christensen and J. Newman. Stress generation and fracture in lithium insertion materials. J. Solid State Electrochem., 10(5):293–319, 2006.

[57] P. Ramadass, Bala Haran, Ralph White, and Branko N. Popov. Mathematical modeling of the capacity fade of Li-ion cells. J. Power Sources, 123(2):230–240, 2003.

[58] P. Ramadass, Bala Haran, Parthasarathy M. Gomadam, Ralph White, and Branko N. Popov. Development of First Principles Capacity Fade Model for Li-Ion Cells. J. Electrochem. Soc., 151(2):A196, 2004.

[59] J. Purewal, J. Wang, J. Graetz, S. Soukiazian, H. Tataria, and M. Verbrugge. Degradation of lithium ion batteries employing graphite negatives and nickel–cobalt–manganese oxide + spinel manganese oxide positives: Part 2, chemical–mechanical degradation model. Journal of Power Sources, 272:1154–1161, 12 2014.

[60] N. Legrand, S. Raël, B. Knosp, M. Hinaje, P. Desprez, and F. Lapicque. Including double-layer capacitance in lithium-ion battery mathematical models. J. Power Sources, 251:370–378, April 2014.

[61] L. Mai, X: Tian, X. Xu, L. Chang, and L. Xu. Nanowire Electrodes for Electrochemical Energy Storage Devices. Chem. Rev., 114(23):11828–11862, 2014.

[62] M. Farkhondeh and C. Delacourt. Mathematical Modeling of Commercial LiFePO4 Electrodes Based on Variable Solid-State Diffusivity. J. Electrochem. Soc., 159(2):A177, 2012.

[63] F. Röder, S. Sonntag, D. Schröder, and U. Krewer. Simulating the impact of particle size distribution on the performance of graphite electrodes in lithium-ion batteries. Energy Technology, pages 1588–1597, 2016.

[64] A. Latz and J. Zausch. Thermal-Electrochemical Lithium-Ion Battery Simulations on Microstructure and Porous Electrode Scale. ECS Trans., 69(1):75–81, 2015.

[65] S. Hein, J. Feinauer, D. Westhoff, I. Manke, V. Schmidt, and A. Latz. Stochastic microstructure modeling and electrochemical simulation of lithium-ion cell anodes in 3D. J. Power Sources, 336:161–171, 2016.

[66] F. Röder, R. D. Braatz, and U. Krewer. Multi-Scale Simulation of Heterogeneous Surface Film Growth Mechanisms in Lithium-Ion Batteries. J. Electrochem. Soc., 164(11):E3335–E3344, 2017.

[67] A. Barré, B. Deguilhem, S. Grolleau, M. Gérard, F. Suard, and D. Riu. A review on lithium-ion battery ageing mechanisms and estimations for automotive applications. J. Power Sources, 241:680–689, 2013.

[68] A. M. Colclasure and R. J. Kee. Thermodynamically consistent modeling of elementary electrochemistry in lithium-ion batteries. Electrochimica Acta, 55:8960–8973, 2010.

[69] M. Heinrich, N. Wolff, N. Harting, V. Laue, F. Röder, S. Seitz, and U. Krewer. Physico-chemical modeling of a lithium-ion battery: An ageing study with electrochemical impedance spectroscopy. Batteries & Supercaps, 2:530–540, 2019.

[70] N. Wolff, N. Harting, M. Heinrich, and U. Krewer. Nonlinear frequency response analysis on lithium-ion batteries: Process identification and differences between transient and steady-state behavior. Electrochimica Acta, 298:788–798, 2019.

[71] C. Meyer, M. Kosfeld, W. Haselrieder, and A. Kwade. Process modeling of the electrode calendering of lithium-ion batteries regarding variation of cathode active materials and mass loadings. Journal of Energy Storage, 18:371–379, 2018.

[72] C. Sangrós Giménez, B. Finke, C. Nowak, C. Schilde, and A. Kwade. Structural and mechanical characterization of lithium-ion battery electrodes via DEM simulations. Advanced Powder Technology, 29:2312–2321, 2018.

[73] G. Lenze, F. Röder, H. Bockholt, W. Haselrieder, A. Kwade, and U. Krewer. Simulation-Supported Analysis of Calendering Impacts on the Performance of Lithium-Ion-Batteries. Journal of The Electrochemical Society, 164(6):A1223–A1233, 2017.

[74] B. Kenney, K. Darcovich, D. D. MacNeil, and I. J. Davidson. Modelling the impact of variations in electrode manufacturing on lithium-ion battery modules. Journal of Power Sources, 213:391–401, 2012.

[75] Y.-H. Chen, C.-W. Wang, G. Liu, X.-Y. Song, V. S. Battaglia, and a. M. Sastry. Selection of Conductive Additives in Li-Ion Battery Cathodes. J. Electrochem. Soc., 154(10):A978, 2007.

[76] Y. H. Chen, C. W. Wang, X. Zhang, and a. M. Sastry. Porous cathode optimization for lithium cells: Ionic and electronic conductivity, capacity, and selection of materials. J. Power Sources, 195(9):2851–2862, 2010.

[77] M. M. Forouzan, C. W. Chao, D. Bustamante, B: Mazzeo, and D. R. Wheeler. Experiment and simulation of the fabrication process of lithium-ion battery cathodes for determining microstructure and mechanical properties. J. Power Sources, 312:172–183, 2016.

[78] C.-W. Wang and A. M. Sastry. Mesoscale Modeling of a Li-Ion Polymer Cell. J. Electrochem. Soc., 154(11):A1035, 2007.

[79] G. Inoue and M. Kawase. Numerical and experimental evaluation of the relationship between porous electrode structure and effective conductivity of ions and electrons in lithium-ion batteries. J. Power Sources, 342:476–488, 2017.

[80] A. Etiemble, N. Besnard, A. Bonnin, J. Adrien, T. Douillard, P. Tran-Van, L. Gautier, J. C. Badot, E. Maire, and B. Lestriez. Multiscale morphological characterization of process induced heterogeneities in blended positive electrodes for lithium–ion batteries. J. Mater. Sci., 52(7):3576–3596, 2017.

[81] T: Danner, M. Singh, S. Hein, J. Kaiser, H. Hahn, and A. Latz. Thick electrodes for Li-ion batteries: A model based analysis. J. Power Sources, 334:191–201, 2016.

[82] T. Carraro, J. Joos, B. Rüger, A. Weber, and E. Ivers-Tiffée. 3D finite element model for reconstructed mixed-conducting cathodes: I. Performance quantification. Electrochim. Acta, 77:315–323, 2012.

[83] M. Ender, J. Joos, T. Carraro, and E. Ivers-Tiffee. Quantitative Characterization of LiFePO4 Cathodes Reconstructed by FIB/SEM Tomography. J. Electrochem. Soc., 159(7):A972–A980, 2012.

[84] J. Ott, B. Volker, Y. Gan, R. McMeeking, and M. Kamlah. A micromechanical model for e ff ective conductivity in granular electrode structures. Acta Mech. Sin., 29:682–698, 2013.

[85] D. Westhoff, J. Feinauer, K. Kuchler, T. Mitsch, I. Manke, S. Hein, A. Latz, and V. Schmidt. Parametric stochastic 3D model for the microstructure of anodes in lithium-ion power cells. Comput. Mater. Sci., 126:453–467, 2017.

[86] C.F. Chen, A. Verma, and P. P. Mukherjee. Probing the Role of Electrode Microstructure in the Lithium-Ion Battery Thermal Behavior. J. Electrochem. Soc., 164(11):E3146–E3158, 2017.

[87] Z. Liu and P. P. Mukherjee. Microstructure Evolution in Lithium-Ion Battery Electrode Processing. J. Electrochem. Soc., 161(8):E3248–E3258, May 2014.

[88] A. C. Ngandjong, A. Rucci, M. Maiza, G. Shukla, J. Vazquez-Arenas, and A. A. Franco. Multiscale Simulation Platform Linking Lithium Ion Battery Electrode Fabrication Process with Performance at the Cell Level. J. Phys. Chem. Lett., 8(23):5966–5972, 2017.

[89] T: Knoche, F. Surek, and G. Reinhart. A Process Model for the Electrolyte Filling of Lithium-ion Batteries. Procedia CIRP, 41:405–410, 2016.

[90] S. Ahmed, P. A. Nelson, and D. W. Dees. Study of a dry room in a battery manufacturing plant using a process model. J. Power Sources, 326:490–497, 2016.

[91] A: Negahban and J. S. Smith. Simulation for manufacturing system design and operation: Literature review and analysis. Journal of Manufacturing Systems, 33(2):241 – 261, 2014.

[92] M. Schönemann. Multiscale Simulation Approach for Battery Production Systems (Sustainable Production, Life Cycle Engineering and Management). Springer, Cham, 2017.

[93] M. Thomitzek, O. Schmidt, F. Röder, U. Krewer, C. Herrmann, and S. Thiede. Simulating Process-Product Interdependencies in Battery Production Systems. Procedia CIRP, 72:346–351, 2018.

[94] Chrissoleon T. Papadopoulos, Jingshan Li, and Michael E.J. O'Kelly. A classification and review of timed markov models of manufacturing systems. Computers & Industrial Engineering, 128:219 – 244, 2019.

[95] V. Laue, F. Röder, and U. Krewer. Practical identifiability of electrochemical p2d-models for lithium-ion batteries (**submitted to**). Journal of Applied Electrochemistry, 2020.

[96] J. C. Forman, S. J. Moura, J. L. Stein, and H. K. Fathy. Genetic identification and fisher identifiability analysis of the Doyle-Fuller-Newman model from experimental cycling of a LiFePO4cell. J. Power Sources, 210:263–275, 2012.

[97] J. Marcicki, M. Canova, A. T. Conlisk, and G. Rizzoni. Design and parametrization analysis of a reduced-order electrochemical model of graphite/LiFePO4 cells for SOC/SOH estimation. J. Power Sources, 237:310–324, 2013.

[98] J. Vazquez-Arenas, L. E. Gimenez, M. Fowler, T. Han, and S. K. Chen. A rapid estimation and sensitivity analysis of parameters describing the behavior of commercial Li-ion batteries including thermal analysis. Energy Convers. Manag., 87:472–482, 2014.

[99] R. Masoudi, T. Uchida, and J. McPhee. Parameter estimation of an electrochemistry-based lithium-ion battery model. J. Power Sources, 291:215–224, 2015.

[100] N. Jin, D. L. Danilov, P. M. J. Van den Hof, and M. C. F. Donkers. Parameter estimation of an electrochemistry-based lithium-ion battery model using a two-step procedure and a parameter sensitivity analysis. Int. J. Energy Res., 42(7):2417–2430, 2018.

[101] D. Dvorak, T. Bauml, A. Holzinger, and H. Popp. A Comprehensive Algorithm for Estimating Lithium-Ion Battery Parameters from Measurements. IEEE Trans. Sustain. Energy, 9(2):771–779, 2018.

[102] H. Chun and S. Han. Electrochemical Model Parameter Estimation of a Lithium-Ion Battery Using a Metaheuristic Algorithm: Cascaded Improved Harmony Search. IFAC-PapersOnLine, 51(28):409–413, 2018.

[103] S. Barcellona and L. Piegari. Lithium ion battery models and parameter identification techniques. Energies, 10(12), 2017.

[104] C. Zhang, K. Li, S. McLoone, and Z. Yang. Battery modelling methods for electric vehicles - A review. 2014 Eur. Control Conf. ECC 2014, pages 2673–2678, 2014.

[105] A. Jokar, B. Rajabloo, M. Désilets, and M. Lacroix. Review of simplified Pseudo-two-Dimensional models of lithium-ion batteries. J. Power Sources, 327:44–55, 2016.

[106] S. Nejad, D. T. Gladwin, and D. a. Stone. A systematic review of lumped-parameter equivalent circuit models for real-time estimation of lithium-ion battery states. J. Power Sources, 316:183–196, 2016.

[107] I. Jarraya, J. Loukil, F. Masmoudi, M. H. Chabchoub, and H. Trabelsi. Modeling and parameters Estimation for Lithium-Ion cells in Electric Drive Vehicle. 2018 15th Int. Multi-Conference Syst. Signals Devices, pages 1128–1132, 2018.

[108] H. P. G. J. Beelen, H. J. Bergveld, and M. C.F. Donkers. On Experiment Design for Parameter Estimation of Equivalent-Circuit Battery Models. 2018 IEEE Conf. Control Technol. Appl. CCTA 2018, (1):1526–1531, 2018.

[109] M. Ecker, T. K. D. Tran, P. Dechent, S. Kabitz, a. Warnecke, and D. U. Sauer. Parameterization of a Physico-Chemical Model of a Lithium-Ion Battery: I. Determination of Parameters. J. Electrochem. Soc., 162(9):A1836–A1848, 2015.

[110] X. Li, K. Pan, G. Fan, R. Lu, C. Zhu, G. Rizzoni, and M. Canova. A physics-based fractional order model and state of energy estimation for lithium ion batteries. Part II: Parameter identification and state of energy estimation for LiFePO4 battery. J. Power Sources, 367:202–213, 2017.

[111] M. Ecker, S. Kabitz, I. Laresgoiti, and D. U. Sauer. Parameterization of a Physico-Chemical Model of a Lithium-Ion Battery: II. Model Validation. J. Electrochem. Soc., 162(9):A1849–A1857, 2015.

[112] J. Schmalstieg and D. U. Sauer. Full Cell Parameterization of a High-Power Lithium-Ion Battery for a Physico-Chemical Model: Part II. Thermal Parameters and Validation. J. Electrochem. Soc., 165(16):A3811–A3819, 2018.

[113] A. M. Bizeray, J. H. Kim, S. R. Duncan, and D. Howey. Identifiability and Parameter Estimation of the Single Particle Lithium-Ion Battery Model. IEEE Trans. Control Syst. Technol., 2018.

[114] A. Sharma and H. K. Fathy. Fisher identifiability analysis for a periodically-excited equivalent-circuit lithium-ion battery model. Proc. Am. Control Conf., pages 274–280, 2014.

[115] D. C. López, G. Wozny, A. Flores-Tlacuahuac, R. Vasquez-Medrano, and V. M. Zavala. A computational framework for identifiability and ill-conditioning analysis of lithium-ion battery models. Ind. Eng. Chem. Res., 55(11):3026–3042, 2016.

[116] X. Lin and a. G. Stefanopoulou. Analytic Bound on Accuracy of Battery State and Parameter Estimation. J. Electrochem. Soc., 162(9):A1879–A1891, 2015.

[117] A. P. Schmidt, M. Bitzer, A. W. Imre, and L. Guzzella. Experiment-driven electrochemical modeling and systematic parameterization for a lithium-ion battery cell. J. Power Sources, 195(15):5071–5080, 2010.

[118] Joel C. Forman, Scott J. Moura, Jeffrey L. Stein, and Hosam K. Fathy. Genetic parameter identification of the Doyle-Fuller-Newman model from experimental cycling of a LiFePO<inf>4</inf> battery. Proc. 2011 Am. Control Conf., pages 362–369, 2011.

[119] A. Pozzi, G. Ciaramella, S. Volkwein, and D. M. Raimondo. Optimal design of experiments for a lithium-ion cell: parameters identification of a single particle model with electrolyte dynamics. Ind. Eng. Chem. Res., 58:1286–1299, 2018.

[120] N. Tian, Y. Wang, J. Chen, and H. Fang. On parameter identification of an equivalent circuit model for lithium-ion batteries. 1st Annu. IEEE Conf. Control Technol. Appl. CCTA 2017, 2017-January:187–192, 2017.

[121] A. I. Pozna, A. Magyar, and K. M. Hangos. Model identification and parameter estimation of lithium ion batteries for diagnostic purposes. 19th Int. Symp. Power Electron. Ee 2017, 2017-December:1–6, 2017.

[122] D. Guo and G. Yang. Parameter Identification Method for Fractional-order Model of Lithium-ion Battery. 2018 IEEE Int. Power Electron. Appl. Conf. Expo., 1(2):1415–1420, 2018.

[123] G. Westermeier, M., Zeilinger, T., Reinhart. Method for Quality Parameter Identification and Classification in Battery Cell Production. In 3rd Int. Electr. Drives Prod. Conf., Nuremberg, 2013.

[124] C. Edouard, M. Petit, C. Forgez, J. Bernard, and R. Revel. Parameter sensitivity analysis of a simplified electrochemical and thermal model for Li-ion batteries aging. J. Power Sources, 325:482–494, 2016.

[125] M. Hadigol, K. Maute, and A. Doostan. On uncertainty quantification of lithium-ion batteries: Application to an $LiC_6/LiCoO_2$ cell. J. Power Sources, 300:507–524, 2015.

[126] S. Santhanagopalan and R. E. White. Quantifying Cell-to-Cell Variations in Lithium Ion Batteries. Int. J. Electrochem., 2012:1–10, 2012.

[127] U. Krewer, F. Röder, E. Harinath, R. D. Braatz, B. Bedürftig, and R. Findeisen. Review—Dynamic Models of Li-Ion Batteries for Diagnosis and Operation: A Review and Perspective. J. Electrochem. Soc., 165(16):A3656–A3673, 2018.

[128] S. Santhanagopalan and R. E. White. Modeling Parametric Uncertainty Using Polynomial Chaos Theory. In ECS Trans., volume 3, pages 243–256. ECS, 2007.

[129] S. Santhanagopalan and R. E. White. Quantifying Cell-to-Cell Variations in Lithium Ion Batteries. Int. J. Electrochem., 2012:1–10, 2012.

[130] A. P. Schmidt, M. Bitzer, A. W. Imre, and L. Guzzella. Experiment-driven electrochemical modeling and systematic parameterization for a lithium-ion battery cell. J. Power Sources, 195(15):5071–5080, 2010.

[131] N. Lin, X. Xie, R. Schenkendorf, and U. Krewer. Efficient global sensitivity analysis of 3d multiphysics model for li-ion batteries. J. Electrochem. Soc., 165:A1169–A1183, 2018.

[132] S. F. Schuster, M. J. Brand, P. Berg, M. Gleissenberger, and A. Jossen. Lithium-ion cell-to-cell variation during battery electric vehicle operation. J. Power Sources, 297:242–251, 2015.

[133] M. Dubarry, C. Truchot, M. Cugnet, B. Y. Liaw, K. Gering, S. Sazhin, D. Jamison, and C. Michelbacher. Evaluation of commercial lithium-ion cells based on composite positive electrode for plug-in hybrid electric vehicle applications. Part I: Initial characterizations. J. Power Sources, 196(23):10328–10335, 2011.

[134] D.-W. Chung, P. R. Shearing, N. P. Brandon, S. J. Harris, and R. E. Garcia. Particle Size Polydispersity in Li-Ion Batteries. J. Electrochem. Soc., 161(3):A422–A430, 2014.

[135] H. Bockholt, W. Haselrieder, and A. Kwade. Intensive Dry and Wet Mixing Influencing the Structural and Electrochemical Properties of Secondary Lithium-Ion Battery Cathodes. ECS Trans., 50(26):25–35, 2013.

[136] H. Bockholt, M. Indrikova, A. Netz, F. Golks, and A. Kwade. The interaction of consecutive process steps in the manufacturing of lithium-ion battery electrodes with regard to structural and electrochemical properties. J. Power Sources, 325:140–151, 2016.

[137] Y.-B. Yi, C.-W. Wang, and A. M. Sastry. Compression of Packed Particulate Systems: Simulations and Experiments in Graphitic Li-ion Anodes. J. Eng. Mater. Technol., 128(1):73, 2006.

[138] C.-W. Wang, Y.-B. Yi, A. M. Sastry, J. Shim, and K. a. Striebel. Particle Compression and Conductivity in Li-Ion Anodes with Graphite Additives. J. Electrochem. Soc., 151(9):A1489, 2004.

[139] H. Zheng, L. Tan, G. Liu, X. Song, and V. S. Battaglia. Calendering effects on the physical and electrochemical properties of Li[Ni1/3Mn1/3Co1/3]O2 cathode. J. Power Sources, 208:52–57, 2012.

[140] B. G. Westphal, H. Bockholt, T. Gunther, W. Haselrieder, and A. Kwade. Influence of Convective Drying Parameters on Electrode Performance and Physical Electrode Properties. ECS Trans., 64(22):57–68, April 2015.

[141] G. W. Tyler. Numerical integration of functions of several variables. Can. Jn. Math., 5:393–412, 1953.

[142] E. Rosenblueth. Point estimates for probability moments. Proc. Natl. Acad. Sci. U S A, 72(10):3812–4, 1975.

[143] Z. Lin and W. Li. Restrictions of point estimate methods and remedy. Reliab. Eng. Syst. Saf., 111:106–111, 2013.

[144] R. Schenkendorf. A General Framework for Uncertainty Propagation Based on Point Estimate Methods. In Proc. 2nd Eur. Conf. Progn. Heal. Manag. Soc. 2014, Fort Worth, 2014.

[145] R. C. Smith. Uncertainty Quantification: Theory, Implementation, and Application. Society for Industrial and Applied Mathematics, Philadelphia, 2014.

[146] N. Legrand, S. Raël, B. Knosp, M. Hinaje, P. Desprez, and F. Lapicque. Including double-layer capacitance in lithium-ion battery models. Journal of Power Sources, 251:370–378, 20014.

[147] V. Ramadesigan, P. W. C. Northrop, S. De, S. Santhanagopalan, R. D. Braatz, and V. R. Subramanian. Modeling and Simulation of Lithium-Ion Batteries from a Systems Engineering Perspective. J. Electrochem. Soc., 159(3):R31–R45, 2012.

[148] S. Mendoza, M. Rothenberger, J. Liu, and H. K. Fathy. Maximizing Parameter Identifiability of a Combined Thermal and Electrochemical Battery Model Via Periodic Current Input Optimization. IFAC-PapersOnLine, 50(1):7314–7320, 2017.

[149] W. M. Haynes. CRC Handbook of Chemistry and Physics. CRC Press, 91 edition, 2011.

[150] M. Dubarry, N. Vuillaume, and B. Y. Liaw. From single cell model to battery pack simulation for Li-ion batteries. J. Power Sources, 186(2):500–507, 2009.

[151] M. Dubarry, A. Devie, and B. Y. Liaw. Cell-balancing currents in parallel strings of a battery system. J. Power Sources, 321:36–46, 2016.

[152] C. von Lüders, V. Zinth, S. V. Erhard, P. J. Osswald, M. Hofmann, R. Gilles, and A. Jossen. Lithium plating in lithium-ion batteries investigated by voltage relaxation and in situ neutron diffraction. J. Power Sources, 342:17–23, 2017.

[153] C. Pastor-Fernández, T. Bruen, W. D. Widanage, M. a. Gama-Valdez, and J. Marco. A Study of Cell-to-Cell Interactions and Degradation in Parallel Strings: Implications for the Battery Management System. J. Power Sources, 329:574–585, 2016.

[154] T. V. Reshetenko, K. Bethune, and R. Rocheleau. Spatial proton exchange membrane fuel cell performance under carbon monoxide poisoning at a low concentration using a segmented cell system. J. Power Sources, 218:412–423, 2012.

[155] T. V. Reshetenko, K. Bethune, M. A. Rubio, and R. Rocheleau. Study of low concentration CO poisoning of Pt anode in a proton exchange membrane fuel cell using spatial electrochemical impedance spectroscopy. J. Power Sources, 269:344–362, 2014.

[156] T. V. Reshetenko and J. St-Pierre. Study of the aromatic hydrocarbons poisoning of platinum cathodes on proton exchange membrane fuel cell spatial performance using a segmented cell system. J. Power Sources, 333:237–246, 2016.

[157] U. Krewer, A. Kamat, and K. Sundmacher. Understanding the dynamic behaviour of direct methanol fuel cells: Response to step changes in cell current. J. Electroanal. Chem., 609(2):105–119, 2007.

[158] S. S. Zhang. The effect of the charging protocol on the cycle life of a Li-ion battery. J. Power Sources, 161(2):1385–1391, 2006.

[159] L. Somerville, J. Bareno, S. Trask, P. Jennings, A. McGordon, C. Lyness, and I. Bloom. The effect of charging rate on the graphite electrode of commercial lithium-ion cells: A post-mortem study. J. Power Sources, 335:189–196, 2016.

[160] J. Schnell, T. Günther, T. Knoche, C. Vieider, L. Köhler, A. Just, M. Keller, S. Passerini, and G. Reinhart. All-solid-state lithium-ion and lithium metal batteries – paving the way to large-scale production. J. Power Sources, 382(February):160–175, 2018.

[161] Y. J. Nam, D. Y. Oh, S. H. Jung, and Y. S. Jung. Toward practical all-solid-state lithium-ion batteries with high energy density and safety: Comparative study for electrodes fabricated by dry- and slurry-mixing processes. J. Power Sources, 375(November 2017):93–101, 2018.

[162] N. Wolff, F. Röder, and U. Krewer. Model Based Assessment of Performance of Lithium-Ion Batteries Using Single Ion Conducting Electrolytes. Electrochimica Acta, 284:639–646, 07 2018.

[163] A. Attari Moghaddam, M. Prat, E. Tsotsas, and A. Kharaghani. Evaporation in Capillary Porous Media at the Perfect Piston-Like Invasion Limit: Evidence of Nonlocal Equilibrium Effects. Water Resour. Res., 53(12):10433–10449, 2017.

[164] R. Amin, P. Balaya, and J. Maier. Anisotropy of Electronic and Ionic Transport in LiFePO4 Single Crystals. Electrochem. Solid-State Lett., 10(1):A13, 2007.

[165] Y.-N. Xu, S.-Y. Chung, J. T. Bloking, Y.-M. Chiang, and W. Y. Ching. Electronic Structure and Electrical Conductivity of Undoped LiFePO[sub 4]. Electrochem. Solid-State Lett., 7(6):A131, 2004.

[166] Y. D. Cho, G. T. K. Fey, and H. M. Kao. The effect of carbon coating thickness on the capacity of LiFePO4/C composite cathodes. J. Power Sources, 189(1):256–262, 2009.

[167] R. Dominko, M. Bele, M. Gaberscek, M. Remskar, D. Hanzel, S. Pejovnik, and J. Jamnik. Impact of the Carbon Coating Thickness on the Electrochemical Performance of LiFePO[sub 4]/C Composites. J. Electrochem. Soc., 152(3):A607, 2005.

[168] M. Grünebaum, M. M. Hiller, S. Jankowsky, S. Jeschke, B. Pohl, T. Schürmann, P. Vettikuzha, A. C. Gentschev, R. Stolina, R. Müller, and H. D. Wiemhöfer. Synthesis and electrochemistry of polymer based electrolytes for lithium batteries. Prog. Solid State Chem., 42(4):85–105, 2014.

[160] A. Polz, M. Grünbaum, and H. D. Wiemhöfer. Hybrid electrolytes for lithium ion and post lithium ion batteries. Encyclopedia of Interfacial Chemistry, pages 660 – 673, 2018.

[170] A. Sakuda, K. Kuratani, M. Yamamoto, M. Takahashi, T. Takeuchi, and H. Kobayashi. All-Solid-State Battery Electrode Sheets Prepared by a Slurry Coating Process. J. Electrochem. Soc., 164(12):A2474–A2478, 2017.

[171] D. E. Stephenson, E. M. Hartman, J. N. Harb, and D. R. Wheeler. Modeling of Particle-Particle Interactions in Porous Cathodes for Lithium-Ion Batteries. J. Electrochem. Soc., 154(12):A1146, 2007.

[172] N. Zacharias, D. R. Nevers, C. Skelton, K. Knackstedt, D. E. Stephenson, and D. R. Wheeler. Direct Measurements of Effective Ionic Transport in Porous Li-Ion Electrodes. J. Electrochem. Soc., 160(2):A306–A311, 2013.

[173] H. Zheng, R. Yang, G. Liu, X. Song, and V. S. Battaglia. Cooperation between active material, polymeric binder and conductive carbon additive in lithium ion battery cathode. J. Phys. Chem., 116(7):4875–4882, 2012.

[174] A. Vadakkepatt, B. Trembacki, S. R. Mathur, and J. Y. Murthy. Bruggeman's exponents for effective thermal conductivity of lithium-ion battery electrodes. Journal of the Electrochemical Society, 163(2):A119–A130, 2016.

[175] A. Bielefeld, D. Weber, and J. Janek. Microstructural Modeling of Composite Cathodes for All-Solid-State Batteries. J. Phys. Chem. C, 123(3):1626–1634, 2019.

[176] H. Dreger, W. Haselrieder, and A. Kwade. Influence of dispersing by extrusion and calendering on the performance of lithium-ion battery electrodes. Journal of Energy Storage, 21:231–240, 2019.

[177] S. Marelli, C. Lamas, B. Sudret, K. Konakli, and C. Mylonas. UQLab user manual – Sensitivity analysis. Technical report, Chair of Risk, Safety & Uncertainty Quantification, ETH Zurich, 2018. Report No. UQLab-V1.1-106.

[178] B. Rumberg, B. Epding, I. Stradtmann, and A. Kwade. Identification of li ion battery cell aging mechanisms by half-cell and full-cell open-circuit-voltage characteristic analysis. Journal of Energy Storage, 25, 2019.

List of Figures

List of Tables